TEACHING ENERGY
ACROSS THE SCIENCES
K–12

Edited by
Jeffrey Nordine

National Science Teachers Association

Arlington, Virginia

Claire Reinburg, Director
Wendy Rubin, Managing Editor
Rachel Ledbetter, Associate Editor
Amanda O'Brien, Associate Editor
Donna Yudkin, Book Acquisitions Coordinator

ART AND DESIGN
Will Thomas Jr., Director
Himabindu Bichali, Graphic Designer, cover and
interior design

PRINTING AND PRODUCTION
Catherine Lorrain, Director

NATIONAL SCIENCE TEACHERS ASSOCIATION
David L. Evans, Executive Director
David Beacom, Publisher

1840 Wilson Blvd., Arlington, VA 22201
www.nsta.org/store
For customer service inquiries, please call 800-277-5300.

NSTA is committed to publishing material that promotes the best in inquiry-based science education. However, conditions of actual use may vary, and the safety procedures and practices described in this book are intended to serve only as a guide. Additional precautionary measures may be required. NSTA and the authors do not warrant or represent that the procedures and practices in this book meet any safety code or standard of federal, state, or local regulations. NSTA and the authors disclaim any liability for personal injury or damage to property arising out of or relating to the use of this book, including any of the recommendations, instructions, or materials contained therein.

Additional applicable standard operating procedures can be found in *NSTA's Safety in the Science Classroom, Laboratory, or Field Sites (www.nsta.org/docs/SafetyInTheScienceClassroomLabAndField.pdf)*. Students should be required to review the document or one similar to it under the direction of the teacher. Both student and parent/guardian should then sign the document acknowledging procedures that must be followed for a safer working/learning experience in the laboratory. Additional safety issues and resources can be located at the NSTA Safety Portal (*www.nsta.org/safety*). Disclaimer: The safety precautions of each activity are based in part on use of the recommended materials and instructions, legal safety standards, and better professional practices. Selection of alternative materials or procedures for these activities may jeopardize the level of safety and therefore is at the user's own risk.

Library of Congress Cataloging-in-Publication Data
Names: Nordine, Jeffrey, editor.
Title: Teaching energy across the sciences, K-12 / edited by Jeffrey Nordine.
Description: Arlington, VA : National Science Teachers Association, [2016] |
  Includes bibliographical references and index.
Identifiers: LCCN 2015041254| ISBN 9781941316016 (print) | ISBN 9781941316375
  (e-book)
Subjects: LCSH: Power resources--Study and teaching.
Classification: LCC TJ163.2 .T36 2016 | DDC 531/.60712--dc23
LC record available at http://lccn.loc.gov/2015041254

TEACHING ENERGY
ACROSS THE SCIENCES
K–12

CONTENTS

SECTION 3

SUPPORTING TEACHERS IN EMPHASIZING ENERGY AS A CROSSCUTTING CONCEPT
171

ACKNOWLEDGMENTS

The foundation for the ideas presented in this book comes from a series of two international summits on the teaching and learning of energy, which were held in 2012 and 2013 and funded by the National Science Foundation (grant NSF DUE-0928666).

The first of these summits sought to clarify what the research community has learned about energy as a crosscutting science concept and to identify trends, challenges, and future research needs in the field of energy education. Participants in this summit included science education researchers, scientists, and science teachers. This first summit resulted in the book *Teaching and Learning of Energy in K–12 Education* (Chen et al. 2014), which provides an overview of what the science education research community understands students ought to know about energy, challenges associated with the teaching and learning of energy, and promising approaches to energy instruction.

The second summit built on the first and was focused on the practice of teaching energy in grades K–12. The majority of participants in the first summit were scientists and science education researchers, with only a small group of teacher-participants; in contrast, the majority of participants in the second summit were teachers, with only a small group of scientists and science education researchers. For continuity, all of the teachers who attended the first summit also attended the second, and all of the scientists and science education researchers who attended the second summit also attended the first.

Researcher-participants in the first summit submitted papers based on their research and teacher-participants in the second summit submitted lesson plans from their teaching; these submissions formed the basis of discussion at both summits. The issues, ideas, and strategies related to the teaching and learning of energy that emerged from these summits form the foundation of this book.

The individuals in Table 1 participated in the first (research-focused) summit.

Table 1.

NAMES AND ORGANIZATIONS OR SCHOOLS OF INDIVIDUALS WHO PARTICIPATED IN THE RESEARCH-FOCUSED STUDY

Name	Organization or School
Charles (Andy) Anderson	Michigan State University
Nicole Becker	Michigan State University
Robert Chen	University of Massachusetts Boston
Costas Constantinou	University of Cyprus, Cyprus

Table 1 (*continued*)

Name	Organization or School
Melanie Cooper	Michigan State University
Jenny Dauer	Michigan State University
George DeBoer	Project 2061, American Association for the Advancement of Science
Reinders Duit	Leibniz Institute for Science and Mathematics Education (IPN), Germany
Arthur Eisenkraft	University of Massachusetts Boston
Orna Fallik	Weizmann Institute of Science, Israel
David Fortus	Weizmann Institute of Science, Israel
Cari Herrmann-Abell	Project 2061, American Association for the Advancement of Science
Hui Jin	Ohio State University
Tom Kim	CREATE for STEM Institute, Michigan State University
Joseph Krajcik	CREATE for STEM Institute, Michigan State University
Sara Lacy	TERC, Massachusetts
Yaron Lehavi	David Yellin Academic College of Education, Weizmann Institute of Science, Israel
Xiufeng Liu	State University of New York at Buffalo
Ramon Lopez	University of Texas at Arlington
Alycia Meriweather	Detroit Public Schools
Robin Millar	University of York, England
Hannah Miller	Michigan State University
Kongju Mun	Michigan State University
Knut Neumann	Leibniz Institute for Science and Mathematics Education (IPN), Germany
Jeffrey Nordine	Leibniz Institute for Science and Mathematics Education (IPN), Germany
Ann Novak	Green Hills School, Michigan
Sebastian Opitz	Leibniz Institute for Science and Mathematics Education (IPN), Germany
Nikos Papadouris	University of Cyprus, Cyprus
Mihwa Park	State University of New York at Buffalo
Pamela Pelletier	Boston Public Schools
Helen Quinn	Stanford University
Wei Rui	Beijing Normal University, China
Allison Scheff	University of Massachusetts Boston
Lane Seeley	Seattle Pacific University
Angelica Stacy	University of California, Berkeley
Roger Tobin	Tufts University
Sonia Underwood	Michigan State University
Margot Vigeant	Bucknell University
Lei Wang	Beijing Normal University, China

Table 1 (*continued*)

Name	Organization or School
Xin Wei	Ohio State University
Kristen Wendell	University of Massachusetts Boston
Holger Wendlandt	Käthe-Kollwitz-Schule, Germany

The individuals in Table 2 participated in the second (practice-focused) summit.

Table 2.

NAMES AND ORGANIZATIONS OR SCHOOLS OF INDIVIDUALS WHO PARTICIPATED IN THE PRACTICE-FOCUSED STUDY

Name	Organization or School
Brenda Breil	P. K. Yonge Developmental Research School, Florida
Amanda Chapman	North East Independent School District, Texas
Robert Chen	University of Massachusetts Boston
Michael Clinchot	Boston Public Schools
Arthur Eisenkraft	University of Massachusetts Boston
Orna Fallik	Central School District and Weizmann Institute of Science, Israel
Katie Fitch	Spring Branch Independent School District, Texas
David Fortus	Weizmann Institute of Science, Israel
Christine Gleason	Greenhills School, Michigan
Robyn Hannigan	University of Massachusetts Boston
Nick Kapura	Boston Public Schools
Joseph Krajcik	Michigan State University
Amy Lazarowicz	Detroit Public Schools
Yaron Lehavi	David Yellin Academic College of Education and Weizmann Institute of Science, Israel
Tatiana Lim-Breitbart	Aspire California College Preparatory Academy
Ramon Lopez	University of Texas at Arlington
Alycia Meriweather	Detroit Public Schools
Mike Metcalfe	Verulam School, England
Knut Neumann	Leibniz Institute for Science and Mathematics Education (IPN), Germany
Jeffrey Nordine	Leibniz Institute for Science and Mathematics Education (IPN), Germany
Ann Novak	Greenhills School, Michigan
Michael Novak	Park View School – District 70, Illinois
Angela Palo	Boston Public Schools
Pamela Pelletier	Boston Public Schools

Table 2 (*continued*)

Name	Organization or School
Helen Quinn	Stanford University
Liz Ratashak	Vicksburg Warren School District, Mississippi
Joy Reynolds	Detroit Public Schools
Allison Scheff	University of Massachusetts Boston
Erica Smith	Cuba-Rushford Central School District, New York
Angelica Stacy	University of California, Berkeley
Gerd Stein	Alfred-Nobel-Schule Geesthacht, Germany
Rob Stevenson	University of Massachusetts Boston
Jennifer Stone	Boston Public Schools
Roberta Tanner	Thompson School District, Colorado
Margot Vigeant	Bucknell University
Wang Weizhen	Second High School, China
Holger Wendlandt	Käthe-Kollwitz-Schule, Germany
Huang Yanning	Capital Normal University, China

Reference

Chen, R. F., A. Eisenkraft, D. Fortus, J. S. Krajcik, K. Neumann, J. C. Nordine, and A. Scheff, eds. 2014. *Teaching and learning of energy in K–12 education.* New York: Springer.

FOREWORD
WHY IS THIS BOOK NEEDED?

HELEN QUINN

In the *Next Generation Science Standards (NGSS)*, there are seven so-called crosscutting concepts, which are advocated as one of the three dimensions of science learning (i.e., science and engineering practices, crosscutting concepts, and disciplinary core ideas). As listed in *A Framework for K–12 Science Education* (NRC 2012, p. 3), these seven concepts are as follows:

1. Patterns

2. Cause and effect: Mechanism and explanation

3. Scale, proportion, and quantity

4. Systems and system models

5. Energy and matter: Flows, cycles, and conservation

6. Structure and function

7. Stability and change

These concepts play a role across all disciplines of science and, yet, are rarely taught explicitly in traditional science curricula. The *NGSS* not only introduce these concepts explicitly, they encourage their use, along with the use of science practices, to build a connective tissue that relates different science concepts and builds student competence in applying them in unfamiliar problem contexts. I see the crosscutting concepts as providing a set of problem-solving perspectives. Each concept suggests ways of looking at a system and asking questions about it that are important to understanding the phenomena that occur (or do not occur) in that system.

This book is about a part of one of those seven crosscutting concepts, Energy and Matter: Flows, Cycles, and Conservation. The name is actually backward because what is important in understanding systems is the recognition that energy and matter are conserved and that their conservation has a major consequence. Understanding where energy and matter come from and where they go—their flows and cycles into, out of, and within a system—provides critical information that helps us understand the functioning of that system more broadly. That functioning is constrained, or limited, by the availability of these two related but distinguishable resources. Energy and matter are linked as a single crosscutting concept because they share the feature of conservation, which can be applied across systems in similar ways; however, there are significant differences in the concepts of energy and matter and the issues related to teaching these concepts.

Thus, it makes sense to devote a book solely to energy and the ways in which its use as a crosscutting concept supports and illuminates science learning across all science disciplines. There is a need for such a book because the ways we have traditionally taught about energy have not stressed the value of looking at energy flows as a tool to understand aspects of a system's behavior, nor have they allowed students a sufficiently deep and broad view of energy to be able to see the connections across disciplines in the application of energy concepts. Yes, ideas about energy are taught in the physical sciences and the term *energy* is used across the science curriculum, but we have not done enough to help students connect energy ideas across disciplines. As the K–12 science curriculum has traditionally been taught, a student would have a great deal of difficulty making any connection between, for example, the way the term is used in a biology class and what they learn about energy in physics class.

This book arose from a set of workshops around the teaching of energy. The authors' conclusions and ideas were developed in the context of those workshops: They seek to address why and how we must approach teaching about energy differently to enable students to use this concept as a tool to address and understand new problems and contexts. I believe that teachers will find these perspectives very useful as they work to help their students understand the crosscutting nature and significance of energy and apply the idea of its conservation to reason about a wide range of systems.

Reference

National Research Council (NRC). 2012. *A framework for K–12 science education: Practices, crosscutting concepts, and core ideas*. Washington, DC: National Academies Press.

SECTION 1

EXPLORING ENERGY AS A CROSSCUTTING CONCEPT

CHAPTER 1

WHY IS ENERGY IMPORTANT?

JEFFREY NORDINE

Conservation of energy is one of the most simply stated principles in all of science—that is, initial energy equals final energy for any isolated system. Yet, some scientists devote their entire career to understanding how energy behaves, and students often have a great deal of difficulty in understanding energy principles and applying them to make sense of everyday systems. Why is this?

Although the simplicity of the energy conservation principle makes it so critically important in science, its application in complex real-world systems can be very difficult indeed. Most real-world systems are not isolated; they allow energy to transfer into and out of them. If a swinging pendulum acted like many science classrooms assume, then it would swing back and forth forever. But this is not the way the world works. Pendulums slow down and eventually stop, clocks need to be wound, and batteries eventually need to be recharged or replaced. Thus, while it is easy to state the conservation principle for any isolated system, it is difficult to find a system that we can reasonably think of as isolated and actually calculate a numerical value of the energy within that system so that it is possible to track changes in energy that occur.

In schools, teachers commonly ask students to use the energy conservation principle to solve problems involving prediction of the behavior of real systems, such as pendulums and model roller coasters. It doesn't take long for students to notice that their answers are always wrong. The more interesting the system (e.g., more complex machines and faster-moving objects), the more their predictions tend to differ from results. Before long, conservation of energy can begin to look like abstract scientific dogma that is really useful only for solving problems from a textbook. One problem with the typical approach to energy instruction in schools is that it often fails to show students how the principle is useful outside of the classroom setting.

Further complicating the issue is the fact that energy ideas pervade our everyday lives and have many nonscientific meanings. Most students show up to school already having a set of intuitive ideas about energy. Prekindergarten students have likely been told by their parents, "Whoa, you have a lot of energy today!" Older students have likely noticed commercials on television that tout an oil company's role in energy production. Virtually all children have been asked to turn off the lights to conserve energy (see Figure 1.1).

Figure 1.1. A sign found next to many light switches

Students' experiences with hearing and using the term *energy* to describe everyday events give them a very intuitive sense of what energy is and how it behaves. Yet, these intuitive feelings are often at odds with school science instruction. After years of seeing batteries die or being asked to turn off the lights to save energy, their science teacher suddenly tells them that energy is *never* used up and that energy is *always* conserved, no matter what they do! When classroom instruction seems to conflict with—rather than clarify—their intuitive ideas about energy, students struggle to develop a strong and self-consistent understanding of the energy concept that is useful for interpreting phenomena and events across in-school and out-of-school contexts (Driver et al. 1994). However, intentionally designed instruction can help ensure that students develop a set of connected ideas that are applicable in a wide range of contexts.

Why Is Energy Learning So Important in Today's World?

The energy concept is fundamentally scientific in nature, but it has tremendous personal and social consequence. Even without explicitly thinking about energy as a concept, people make energy-related decisions every day. For example, as people make dietary decisions, a key consideration is that they ingest an appropriate number of calories each day to carry on life processes without exceeding this amount. Said another way, we need to be sure that the energy inputs (calories we eat) and outputs (calories we "burn") for our human body system are in balance. It requires very little scientific or mathematical sophistication to engage in the process of counting calories. An individual needs only to track the number of calories that he or she ingests and to approximate the daily calories burned through his or her activities—a process that has been made exceptionally simple by a variety of smartphone applications. If people ingest more calories than they burn, they will almost certainly gain weight over time. If they ingest fewer than they burn, they will almost certainly lose weight over time.[1] Most people can use calorie tracking to explain changes in their weight, but few deeply understand the complex chemical processes of metabolism that link the energy associated with the arrangement of atoms in our food to the change in mass that manifests in our bodies as a result of a calorie imbalance. The beauty of the energy concept is that a person can use it to predict and explain complex phenomena without needing to understand the full details of the human body system.

The human body houses some of the most complex systems in the known universe, and it is remarkable that a few simple ideas about energy can help us predict and explain how it will respond to changes. Most modern technological systems and devices are also enormously complex, and people make decisions every day that affect how they will function. Whether considering how to lower monthly electric bills, extend the battery life on a cell phone, or improve the fuel efficiency of a vehicle, people can make energy-related decisions without a full understanding of the mechanical or electrical processes that are going on. However, not all energy-related decisions are equally easy to make. For example, most people know that using an air conditioner to make a house very cold on a hot day will lead to higher electric bills, but it is typically less clear what settings have the biggest effect on the length of time that a cell phone battery will last. Lowering energy bills and extending cell phone battery life may feel like fundamentally personal decisions, but in today's interconnected world, these choices affect broader systems. The cumulative effects of people's everyday energy-related decisions have significant consequences for how societies function and the natural environment.

Although many people make energy-related decisions without a sophisticated scientific conception of energy, a deep understanding of energy can help us know the right questions to ask as we make energy-related decisions or evaluate energy-related claims. For example, too many people are fooled by nutrition charlatans who claim that they have a strategy to "Eat all you want and still lose weight!" Similarly, too few people think to ask how the electricity for a zero-emissions electric car was generated in the first place; if it was generated at a coal-fired power plant, it just means that the emissions happened at the power plant instead of the car. Many important personal and social decisions we have to make in the world today are related to energy. If people hold a deep understanding of energy, they are less susceptible to being fooled by energy-related claims that are too good to be true. It turns out, though, that far too many students leave school without understanding some of the most important ideas about energy and how it affects their lives (Liu and McKeough 2005; Neumann et al. 2013).

---

1 Scientists have learned an incredible amount about weight management in recent years and have found that weight gain or loss can vary widely in individuals based on factors such as genetics, type of food, sleep patterns, and environment.

What Makes Learning About Energy Particularly Challenging?

Even as early as prekindergarten, students have developed a set of intuitive ideas about how the world works. Virtually all students have heard and used the term *energy* in a variety of ways to describe a wide range of everyday events well before they learn about energy in school. It is perhaps no surprise, then, that even very young students enter school with a set of intuitive ideas about energy and that these ideas often contradict one another.

Among the most common alternative conceptions that students form about energy is that it is primarily associated only with living things (Solomon 1983; Watts 1983). At the same time, students may hold a range of other ideas about energy, such as that it is associated only with obvious activity (e.g., movement or burning), that it is fundamentally connected to technical devices, or that it is a substance that flows from one place to another (e.g., around an electric circuit) (Domenéch et al. 2007; Driver et al. 1994; Solomon 1983; Trumper 1998). Though it may not seem logical that a student could think of energy as primarily associated with living things and simultaneously hold the belief that it is primarily connected to technical devices, students commonly fail to notice disharmony among their ideas because they are so strongly connected to the contexts in which they use them. That is, student ideas tend to be situated within particular contexts (Lave and Wenger 1992), and certain ideas are more strongly cued by some events than others (diSessa 1993). The notion of situated cognition helps explain why students can simultaneously hold contradictory ideas about energy, but it provides little insight about why energy itself is a difficult concept to learn.

Although we experience some notion of energy every day, the concept can be very difficult to define. Many textbooks offer definitions such as "energy is the ability to do work" or "energy is the capacity to cause a change," but these definitions are often circular (in the first example, "work" is an energy transfer process measured in the same units as energy) or so broad that their utility must be questioned (in the second example, simply associating energy with change does little to nail down what it is as a scientific idea). Energy is a fundamentally abstract concept that eludes a clean definition. In his famous lectures on physics, Richard Feynman (a Nobel Laureate in physics) said,

> It is important to realize that in physics today, we have no knowledge of what energy is. We do not have a picture that energy comes in little blobs of a definite amount. It is not that way. However, there are formulas for calculating some numerical quantity, and when we add it all together it gives "28"— always the same number. It is an abstract thing in that it does not tell us the mechanism or the reasons for the various formulas. (Feynman, Leighton, and Sands 1989, p. 4-2)

So, energy is a ubiquitous concept in our world and is a quantity that we can calculate very precisely, yet it is very difficult to define in specific and useful terms. A useful analogy here is the concept of time. Most of us have a very good sense of how long a minute is. We can measure time very precisely. We use it every day, and most schools run on a very precisely timed schedule. But what *is* time? What is it that a ticking clock is actually measuring? Nearly all of us rely on clocks in our everyday lives and use time to coordinate events with others, yet a formal definition is almost never the topic of conversation. Like time, energy is ubiquitous in our everyday lives and we use our intuitive notions of it to explain what we see and feel; yet when we try to formalize the concept in science class, the formalized definitions may conflict with what students have used in so many different ways outside of the school walls.

To further complicate the issue, the way energy is presented in school tends to vary from subject to subject! When studying life science, students often study energy flow through ecosystems and may learn that only 10% of energy is transferred between trophic levels. In physical science, students commonly assume that 100% of energy is conserved within a system and use this assumption to perform calculations or make predictions about the system's behavior. When studying chemistry in high school, students may use energy to predict and explain chemical reactions by calculating changes in enthalpy (a very specific way to calculate energy). When studying Earth science, students may learn that rocks along fault lines can store energy that gets released during earthquakes. Why have we been presenting energy so differently in different scientific contexts? Because scientists use energy in different ways, and they do so for good reasons.

What Is the Energy Concept, and How Do Scientists Use It?

The scientific concept of energy first grew out of an effort to make sense of objects in motion—a topic of study since antiquity. Scientists first used ideas that we now recognize as related to *kinetic energy* (though they did not initially use this term) in the 17th century to explain the movement of objects. Then, much later, they introduced the notion of *potential energy* (though again, scientists did not initially use this term) to explain how particles interact (Coopersmith 2010). In fact, it wasn't until about halfway through the 19th century that scientists first formulated what we now know as the conservation of energy! It's hard to believe that only about 50 years prior to Einstein publishing his theory of relativity, scientists had not even formulated the law of conservation of energy. So, what took them so long?

As Feynman noted, energy is an inherently abstract quantity that can never be seen nor measured directly. That is, it can only be *calculated* based on the measurement of observable characteristics such as speed, mass, and temperature. The notion of energy arose as a method to understand how changes in the observable characteristics of a system were related. By speculating that something called energy was conserved during interactions, scientists could reliably predict and explain what changes in a system were allowed and which were not. It is important to note that the scientific concept of energy arose because scientists invented an idea—that some unseen quantity must be conserved during any interaction—and tested the logical outcomes of this idea relative to what we observe in nature. In fact, conservation of energy is an assumption about nature that has strikingly little empirical support—in most systems, we are almost always missing some energy that has been transferred to the surroundings (Quinn 2014). Still, we have never observed evidence that the assumption of energy conservation has been violated.

Scientists use the energy concept to make sense of an incredible range of systems, from understanding subatomic interactions between the smallest particles to discerning the structure of the entire known universe. They use it to understand microorganisms that exist without access to sunlight, to engineer the latest smart watch, and to plan voyages to other planets. Although energy principles are applicable everywhere, the conceptual and analytical tools that scientists use to employ energy ideas can look very different indeed.

The fact that energy can look quite different when used as different conceptual and analytical tools may seem strange, but energy is really not so unusual in this respect. If you have ever tried to fix something around the house, you have probably experienced the need for different tools to perform the same basic task in different situations. If, for example, you need to twist something to tighten or loosen it but can't turn it by hand—you need a wrench! A wrench, of course, is a tool designed to help you make something twist. All of the objects in Figure 1.2 are designed to help you twist something that you cannot twist with your bare hands. But even though they are designed to perform the same basic task, these tools look very different from one another. Why? Because they are designed for use in particular situations so that they make twisting as easy as possible in that context.

Just as people have designed lots of different wrenches to make it as easy as possible to twist things in different situations, scientists have developed lots of different tools to make it as easy as possible to apply energy ideas in a wide variety of contexts. Scientists call these tools heat, enthalpy, mechanical work, and so on. Though it may not seem on the surface that the calorie content of the food you eat is fundamentally the same thing as the kinetic energy of a moving roller coaster, it is! When calculating "calories in" and "calories out" via exercise during the course of a person's day, a dietician is tracking energy changes in the human body system. When calculating the kinetic and potential energy of a roller coaster as it moves along the track, a physicist is tracking energy changes in the roller coaster system. The important thing to realize is that the energy tracked by the dietician and the energy tracked by the physicist are *exactly the same thing*. The calories we get from food are ultimately determined by chemical reactions that occur within our bodies as we digest and metabolize it. The energy released during these chemical reactions depends on the kinetic and potential energy changes of the atoms involved in the chemical reactions. But it would be foolish to try to calculate the kinetic energy and potential energy changes for every single atom in our food as we digest and metabolize it! So, scientists invented a method to track energy changes when we eat food without having to account for changes in the energy of individual atoms. Likewise, it wouldn't make any sense to use the idea of calorie content to track the energy changes for a moving roller coaster! The physicist and the dietician use different tools for calculating and tracking energy changes because they are the easiest to use in each context.

Figure 1.2. A variety of objects designed to twist things. Although they are all designed to perform the same basic task, they look very different because they are designed for use in different contexts.

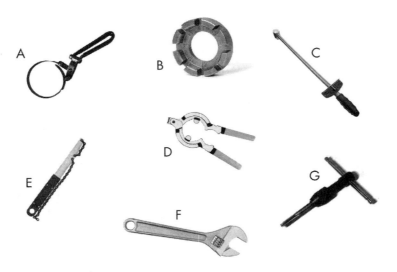

Source: Images A–F, Thinkstock; Image G, Tap Wrench: *https://commons.wikimedia.org/wiki/File:Tap_and_T-wrench.jpg*

When students encounter energy in school, they often fail to see connections among all of the different tools they use to track energy and may come to believe that energy behaves differently in different contexts. There is hope, however. Recent studies have shown that many students do progress toward a more unified and integrated understanding of energy during their time in school (Liu and McKeough 2005; Neumann et al. 2013) and that classroom instruction plays a major role in helping students develop a set of integrated ideas about energy that span contexts and lay the foundation for future learning (Nordine, Krajcik, and Fortus 2011).

This book is intended to unpack why energy is such a critical concept in science and to share successful approaches to energy instruction that exemplify the approach recommended by the *Next Generation Science Standards* (*NGSS*; NGSS Lead States 2013).

Energy in the *NGSS*

The *NGSS* have been redefining a vision of effective science teaching and learning. These standards are a response to our growing understanding of how students learn (Commission on Behavioral and Social Sciences and Education 2000) and the need to provide students with science education that prepares them to function as informed citizens in an increasingly scientific and technological world. The *NGSS* focus on three major dimensions of science learning: scientific and engineering practices, crosscutting concepts, and disciplinary core ideas. Further, they describe how students should build understandings over time through coherent instruction that weaves the three dimensions together—what we call *three-dimensional learning* (NRC 2014). Energy occupies a unique space within the *NGSS* because it is the only major scientific construct that appears as both a disciplinary core idea and a crosscutting concept.

Disciplinary core ideas represent key organizing concepts within disciplines, which means that these ideas can be used to explain a variety of natural phenomena, have social relevance, span grades K–12 (and beyond), and can serve as a foundation for future learning. By focusing on core ideas, instruction becomes more coherent (Roseman, Linn, and Koppal 2008). Coherent instruction aligns learning goals, scientific practices, inquiry tasks, and assessments by focusing on the most central scientific principles (Fortus and Krajcik 2012), and a growing body of evidence suggests that coherent instruction helps students to develop integrated understandings (Lee and Liu 2010; Linn and Eylon 2000; Nordine, Krajcik, and Fortus 2011). If students possess integrated understandings, they are capable of connecting ideas to one another in a relational network that is useful for making sense of relevant phenomena (Linn et al. 2006); further, a growing body of evidence suggests that students with integrated understandings of energy become better prepared for future learning (Nordine and Drake 2012; Nordine, Krajcik, and Fortus 2011). The *NGSS* emphasize coherence in energy instruction by focusing on a small set of core energy ideas, connecting to science and engineering practices, and emphasizing inquiry tasks that span across traditional disciplines.

A Framework for K–12 Science Education (NRC 2012), which serves as the conceptual foundation for the *NGSS*, specifies four major dimensions relating to energy as a disciplinary core idea:

- **PS3.A:** Definitions of Energy
- **PS3.B:** Conservation of Energy and Energy Transfer
- **PS3.C:** Relationship Between Energy and Forces
- **PS3.D:** Energy in Chemical Processes and Everyday Life

For each of these dimensions, the *Framework* and *NGSS* focus on how students should use their understanding to engage in scientific practices as they make sense of relevant and meaningful phenomena. Further, *NGSS* performance expectations are organized such that ideas build on one another over time and new ideas are developed based on existing understandings and the emergence of new evidence—this is the idea of a *learning progression* (Stevens, Delgado, and Krajcik 2009).

The *NGSS* suggest learning progressions for energy by specifying how students should engage in more sophisticated reasoning and analysis over time. For example, in fourth grade, students should begin exploring energy as an explicit concept and be able to "use evidence to construct an explanation relating the speed of an object to the energy of that object" (4-PS3-1). As students grow in sophistication during middle school, they should become more familiar with the energy idea in order to "construct and interpret graphical displays of data to describe the relationships of kinetic energy to the mass of an object and to the speed of an object" (MS-PS3-1). By the end of high school, students should be able to "create a computational model to calculate the change in the energy of one component in a system when the change in energy of the other component(s) and energy flows in and out of the system are known" (HS-PS3-1). Over time, students become more familiar with energy as a construct as they recognize that energy and speed are related (elementary school), identify variables that affect the magnitude of changes in kinetic energy (middle school), and compute numerical values for energy to quantify energy transfers into and out of systems (high school). In conjunction with developing ideas about energy over time, the *NGSS* also emphasize more sophisticated scientific practices by identifying evidence for energy (elementary school), creating representations of data (middle school), and constructing computational models (high school).

Besides being one of a relatively small set of disciplinary core ideas, energy is unusual among the disciplinary core ideas because it is also identified as a crosscutting concept. Crosscutting concepts span disciplinary boundaries to help scientists make sense of phenomena. The crosscutting concepts in the *NGSS* are as follows:

- Patterns
- Cause and effect: Mechanism and explanation
- Scale, proportion, and quantity
- Systems and system models
- Energy and matter: Flows, cycles, and conservation
- Structure and function
- Stability and change

Each crosscutting concept is widely used within each scientific discipline and includes a consistent set of ideas that can be applied across disciplinary boundaries to interpret a broad range of natural phenomena. Energy was first formalized as the scientific principle we know today in the context of the physical sciences (which is why it appears in the *NGSS* as a disciplinary core idea) but, over time, has proved to be an indispensable concept in every scientific discipline (which is why it appears as a crosscutting concept). Biologists use energy to interpret relationships between organisms in an ecosystem; chemists predict and measure changes in chemical reactions using an energy framework; environmental scientists track energy flows through natural and designed systems; physicists use energy to set parameters on the characteristics of yet-undiscovered particles.

Although the energy tools that scientists use in these different contexts can look very different from one another, the *NGSS* firmly assert that we can no longer accept teaching energy in a way that does not show students how energy ideas are connected across scientific disciplines. For students to develop an integrated understanding of energy that applies both inside and outside of school, we must clearly show students how energy can provide a powerful perspective from which they can reliably and consistently view phenomena from any scientific context or discipline. By developing a firm understanding of how to use energy as a perspective for interpreting phenomena, students will be well positioned to extend their learning across time, between school classes, and beyond school walls.

Although the *NGSS* provide a robust set of recommendations for teaching energy in today's schools, they do not provide specific pedagogical approaches or advocate for particular instructional materials.

Strategies for Teaching Energy in Light of the *NGSS*

In an attempt to clarify instructional strategies and approaches that respond to the *NGSS*, we gathered a group of teachers, science educators, and scientists in a series of two international summits on the teaching and learning of energy. During these summits, teachers shared lesson ideas and clarified insights for teaching energy in grades K–12 that exemplify the recommendations in the *NGSS*. We asked teachers to bring lesson plans that focused on one of three learning goals specified within the *Framework*:

1. Many organisms use the energy from light to make sugars (food) from carbon dioxide from the atmosphere and water through the process of photosynthesis, which also releases oxygen. In most animals and plants, oxygen reacts with carbon containing molecules (sugars) to provide energy and produce waste carbon dioxide. (LS1.C)

2. That there is a single quantity called energy is due to the remarkable fact that a system's total energy is conserved as smaller quantities of energy are

transferred between subsystems—or into and out of the system through diverse mechanisms and stored in various ways. (PS3.A)

3. All materials, energy, and fuels that humans use are derived from natural sources, and their use affects the environment in multiple ways. Some resources are renewable over time, and others are not. (ESS3.A)

These learning goals were chosen because they represent three ideas about energy that are easily applied in both scientific and nonscientific settings. Also, these learning goals span the traditional disciplines of school science (i.e., biology, chemistry, physics, Earth science). In discussing approaches to teaching these three learning goals, teachers began to clarify a set of common ideas and phenomena that were useful for communicating the "big ideas" about energy across disciplinary contexts.

In this book, we share a set of ideas that surfaced as a result of this summit and that we refer to as the Five Big Ideas. These ideas can help students think about energy-related phenomena in a consistent way across disciplines. We illustrate how teachers from elementary school through high school can help students develop an ever-increasing understanding of energy by connecting the energy tools used by different disciplines to a consistent set of Five Big Ideas about energy:

*   **Big Idea 1.** All energy is fundamentally the same, and it can be manifested in different phenomena that are often referred to as different "forms" or "types."
*   **Big Idea 2.** Energy can be transformed/converted from one form/type to another.
*   **Big Idea 3.** Energy can be transferred between systems and objects.
*   **Big Idea 4.** Energy is conserved. It is never created or destroyed, only transformed/ converted or transferred.
*   **Big Idea 5.** Energy is dissipated in all macroscopic (involving more than just a few particles) processes.

These Five Big Ideas can be applied consistently across all scientific phenomena, and in this book we will suggest strategies for tuning energy instruction to emphasize these ideas. These Big Ideas can clarify—rather than complicate—your existing energy instruction. By keeping these ideas in mind when designing energy instruction, teachers can put students in a much better position to understand the crosscutting nature of the energy concept and provide them with a consistent lens through which to interpret energy-related contexts that they encounter both in school and their everyday lives.

Organization of This Book

This book is divided into three sections. Section 1 (which includes this chapter) is dedicated to unpacking the scientific concept of energy, identifying what students should know about the concept, and illustrating how changing our perspective on existing energy instruction can bring current activities in line with the recommendations of the *NGSS*.

Section 2 presents approaches to teaching energy. The chapters in this section share insights from the two international summits on energy in which scientists, science educators, and classroom teachers gathered to discuss key issues associated with energy instruction and to specify promising approaches within and across grade levels and disciplines. Section 2 addresses issues of vocabulary use when talking about energy at various grade levels, discusses exemplary approaches for teaching about three key learning goals from the *NGSS* (photosynthesis, conservation, and natural resources), and identifies promising approaches to classroom energy assessment.

Section 3 is primarily written for those who support classroom teachers. The first chapter in this section provides recommendations about offering high-quality professional development for energy teaching and fostering teacher leadership to implement the recommendations within *NGSS*. The final chapter in the book is devoted to instructional support personnel and policy makers who are responsible for the structural components of schooling and providing professional learning opportunities for teachers.

Our hope is that this book will primarily serve as a resource for classroom teachers, and we also hope that it will spur conversation among a range of educators who are responding to the instructional imperatives described by the *Next Generation Science Standards*.

References

Commission on Behavioral and Social Sciences and Education (CBASSE). 2000. *How people learn: Brain, mind, experience, and school: Expanded edition.* Washington, DC: National Academies Press.

Coopersmith, J. 2010. *Energy, the subtle concept: The discovery of Feynman's blocks from Leibniz to Einstein.* New York: Oxford University Press.

diSessa, A. 1993. Toward an epistemology of physics. *Cognition and Instruction* 12 (2–3): 105–225.

Doménech, J. L., D. Gil-Pérez, A. Gras-Martí, J. Guisasola, J. Martínez-Torregrosa, J. Salinas, R. Trumper, P. Valdés, and A. Vilches. 2007. Teaching of energy issues: A debate proposal for a global reorientation. *Science & Education* 16: 43–64.

Driver, R., A. Squires, P. Rushworth, and V. Wood-Robinson. 1994. *Making sense of secondary science: Research into children's ideas.* New York: Routledge.

Feynman, R. P., R. B. Leighton, and M. L. Sands. 1989. *The Feynman lectures on physics.* Vol. 1. Redwood City, CA: Addison-Wesley.

Fortus, D., and J. Krajcik. 2012. Curriculum coherence and learning progressions. In *Second international handbook of science education,* ed. B. J. Fraser, K. Tobin, and C. J.

McRobbie, 783–798. Dordrecht, The Netherlands: Springer. *www.springerlink.com/ index/10.1007/978-1-4020-9041-7.*

Lave, J., and E. Wenger. 1992. *Situated learning: Legitimate peripheral participation.* Cambridge, UK: Cambridge University Press.

Lee, H.-S., and O. L. Liu. 2010. Assessing learning progression of energy concepts across middle school grades: The knowledge integration perspective. *Science Education* 94 (4): 665–688.

Linn, M. C., and B.-S. Eylon. 2000. Knowledge integration and displaced volume. *Journal of Science Education and Technology* 9 (4): 287–310.

Linn, M. C., H.-S. Lee, R. Tinker, F. Husic, and J. L. Chiu. 2006. Teaching and assessing knowledge integration in science. *Science* 313: 1049–1050.

Liu, X., and A. McKeough. 2005. Developmental growth in students' concept of energy: Analysis of selected items from the TIMSS database. *Journal of Research in Science Teaching* 42 (5): 493–517.

National Research Council (NRC). 2012. *A framework for K–12 science education: Practices, crosscutting concepts, and core ideas.* Washington, DC: National Academies Press.

National Research Council (NRC). 2014. *Developing assessments for the* Next Generation Science Standards. Washington, DC: National Academies Press.

Neumann, K., T. Viering, W. J. Boone, and H. E. Fischer. 2013. Towards a learning progression of energy. *Journal of Research in Science Teaching* 50 (2): 162–188. *http://doi.org/10.1002/tea.21061.*

NGSS Lead States. 2013. *Next Generation Science Standards: For states, by states.* Washington, DC: National Academies Press. *www.nextgenscience.org/next-generation-science-standards.*

Nordine, J., and A. Drake, A. 2012. Exploring the relationship between integrated understanding of energy and preparation for future learning. Paper presented at the National Association for Research in Science Teaching (NARST) Annual International Conference, Indianapolis, IN.

Nordine, J., J. Krajcik, and D. Fortus. 2011. Transforming energy instruction in middle school to support integrated understanding and future learning. *Science Education* 95 (4): 670–699. *http:// doi.org/10.1002/sce.20423.*

Quinn, H. 2014. A physicist's musings on teaching about energy. In *Teaching and learning of energy in K–12 education,* eds. R. F. Chen, A. Eisenkraft, D. Fortus, J. S. Krajcik, K. Neumann, J. C. Nordine, and A. Scheff, 15–36. New York: Springer.

Roseman, J. E., M. C. Linn, and M. Koppal. 2008. Characterizing curriculm coherence. In *Designing coherent science education: Implications for curriculum, instruction, and policy,* eds. Y. Kali, M. C. Linn, and J. E. Roseman, 13–36. New York: Teachers College Press.

Solomon, J. 1983. Messy, contradictory, and obstinately persistent: A study of children's out-of-school ideas about energy. *School Science Review* 65 (231): 225–229.

Stevens, S. Y., C. Delgado, and J. S. Krajcik. 2009. Developing a hypothetical multi-dimensional learning progression for the nature of matter. *Journal of Research in Science Teaching* 47 (6): 687–715. *http://doi.org/10.1002/tea.20324.*

Trumper, R. 1998. A longitudinal study of physics students' conceptions of energy in pre-service training for high school teachers. *Journal of Science Education and Technology* 7 (4): 311–317.

Watts, M. 1983. Some alternative views of energy. *Physics Education* 18: 213–217.

CHAPTER 2

WHAT SHOULD STUDENTS KNOW ABOUT ENERGY?

JEFFREY NORDINE

The single most important and remarkable fact about energy is that it is conserved. Whenever the energy of a system or object seems to increase, that energy must have come from somewhere else. Whenever energy seems to decrease, it must go somewhere. Knowing that energy never appears nor vanishes gives scientists a critical clue for interpreting even the most bizarre phenomena. The discovery of the "neutrino" particle is a famous example of how scientists have relied on the energy conservation principle to understand a previously unknown feature of nature. Neutrinos are extremely difficult to detect. They can (and almost always do) pass through the entire Earth without interacting with anything, and trillions pass harmlessly through your body each second. Physicists got a clue to the existence of neutrinos from the fact that no matter how hard they looked, they couldn't find the energy that was missing in a particular type of radioactive decay. It turned out that the missing energy was carried away by neutrinos escaping the system. The energy conservation principle gave physicists a powerful clue that helped them predict and ultimately detect the extremely elusive particle. Scientists' beliefs about energy allow them to discover and understand parts of nature that would be difficult to comprehend without it.

Scientists are not the only ones who develop a strong set of beliefs about energy. We frequently encounter the energy conservation principle in our everyday lives, and throughout these experiences we develop a strong set of intuitions about how energy behaves. We even develop a sense of energy as a conserved quantity, even though we would probably never phrase it like that. Imagine how surprised you'd be if you dropped a marble on the floor and it bounced up and hit the ceiling. Or what if you ordered a cup of coffee and it got *warmer* as you drank it? Although we might not formally define conservation of energy in our everyday lives, we have a sense that we don't get energy for free. We begin to understand that energy doesn't come from nowhere.

Although our everyday experiences help us accept that energy doesn't suddenly appear out of nowhere, there tends to be little to refute the idea that energy cannot simply vanish. We watch a bouncing marble come to rest without ever questioning where its motion energy went; we accept that it requires energy (or a really good insulated cup) to keep coffee hot, but we seldom think about where the energy goes as it cools off. Many times, we think of energy as being stored in something—such as gasoline or batteries—and then simply getting "used up" when a car moves or a flashlight shines. While this idea typically does very well to explain and predict the energy-related situations we encounter in our day-to-day lives, it contradicts the most fundamental and important fact about energy—that it is conserved!

So, we have an interesting situation. Energy is only useful as a scientific idea because it is conserved (i.e., it is never created or destroyed), but many of us learn to successfully predict and explain energy-related scenarios based on the assumption that it is used up and no longer exists—that it is *not* conserved! Most of us learn to reason about energy in our everyday lives based on a fundamental contradiction of what energy is as a scientific concept. Is it any wonder that kids have such a hard time truly accepting the idea of energy conservation (Neumann et al. 2013)?

When science ideas blatantly contradict our everyday experience, they can be hard to accept (Solomon 1983). While many young learners can repeat the mantra that "energy is never created or destroyed," they often simultaneously harbor and activate intuitions that imply the opposite (Chabalengula, Sanders, and Mumba 2012). If we are to be truly successful as teachers of science and help students understand the energy construct, it is critical to understand and identify how learners often think about energy based on their everyday intuitions. After all, if you are a fourth-grade teacher teaching the *Next Generation Science Standards* (*NGSS*; NGSS Lead States 2013), you are responsible for students' first explicit introduction to energy as a scientific idea (even though you won't formally define it yet—keep reading!), and students in your class have had almost 10 years of building their own intuitive ideas about energy in their experiences outside of school.

This chapter is about learning to identify those alternative ideas about energy that students bring into the science classroom and understanding what the *NGSS* suggest about what students ought to know about energy and when they ought to know it.

Alternative Conceptions About Energy

Before we begin thinking about all the so-called wrong ways that kids think about energy, it is essential to remember that their ideas tend to work very well to explain the vast majority of situations. Nobody explicitly taught your students to develop alternative ideas about energy; learners develop intuitive ideas through perhaps their most powerful teaching environment—their experience.

Teaching students about new ideas is not about systematically squashing their existing intuitive ideas. In fact, students build new ideas based on their existing intuition, and for students to learn effectively they must be able to connect their intuitive notions to instructed ideas (Smith, diSessa, and Roschelle 1993). Teaching students to develop more appropriate ideas about energy (or any concept for that matter) is not a find-and-replace operation; rather, our job is to help students connect intuitive and instructed ideas through intentional experiences with appropriate phenomena. By understanding how students' intuitive ideas are related to scientific notions, we can be better prepared to help children identify the limits of their intuitive ideas and develop a more sophisticated understanding of energy over time.

As children begin to develop their intuitive ideas about energy, they extract patterns from their interactions with nature and the ways in which they hear the word *energy* used. Despite the rich variety in children's geographic location and culture, researchers have found that there is remarkable similarity in how students conceptualize the energy concept (Duit 2014). Michael Watts (1983) found that most students' ideas about energy could be categorized into six major categories, shown in Table 2.1.

Table 2.1.

SOME COMMON ALTERNATIVE CONCEPTIONS OF ENERGY

Common alternative conception	Description
Human-centered model	Energy is mainly associated with human beings.
Depository model	Some objects "have" energy, and some "need" it.
Ingredient model	Energy is dormant within some objects or substances, and can be released by some trigger.
Obvious activity model	Energy is identified by overt displays, and the display itself is actually called energy.
Product model	Energy is a by-product of some situation and is relatively short-lived.
Functional model	Energy is a very general kind of fuel, more or less restricted to technical devices and not essential to all processes.
Flow-transfer model	Energy is some sort of physical fluid that is transferred in certain processes.

Source: Adapted from Watts 1983.

When children first begin to develop their own ideas about energy, they typically begin to form ideas associated with their own sense of feeling energetic and then extend that sense to other living things; a human-centered view of energy (that may include other living organisms) is among the most commonly held conceptions among young children. With time and experience, children begin to develop a sense that energy is associated with activity and can be stored or transferred (Driver et al. 1994). It is not until much later—perhaps not even during their high school years—that students commonly display an understanding of energy conservation or degradation (Driver et al. 1994; Liu and McKeough 2005; Neumann et al. 2013).

As students develop their ideas, some commonly held intuitive ideas could be more problematic than others. If, for example, a student believes that energy is only associated with living organisms, then he or she may have trouble understanding the role that energy plays in melting ice. On the other hand, conceptualizing energy as a physical fluid that can flow between objects and systems may present fewer problems as students reason about the role of energy in phenomena (Nordine, Krajcik, and Fortus 2011).

The *NGSS* correspond with what is known about how students tend to build energy ideas over time. Although the word *energy* appears in the *NGSS* as early as kindergarten, energy-related ideas are not explicitly introduced until fourth grade when students are asked to connect their observations to changes in energy, and it is not until middle school that students are asked to use energy ideas to model the behavior of systems. In the next section, we discuss what the *NGSS* say about what students should know about energy and when they should know it.

Energy in the *NGSS*

Energy and matter are the only two ideas in the *NGSS* that appear as both disciplinary core ideas (DCIs) and crosscutting concepts (CCCs). Why is this? Is there a physical science meaning of energy and then a different meaning that applies across all disciplines? No. Energy appears as a DCI because it describes the *physical* interactions in systems, and it appears as a CCC because energy ideas have proved useful as a perspective for analysis across all systems: living, nonliving, natural, and human-designed (Miller 2014).[1] Regardless of the system, every interaction is fundamentally a physical one. Every cellular membrane, tectonic plate, or electric guitar is made of atoms and molecules that interact by exchanging particles (e.g., such as electrons or neutrons) and waves (e.g., light or radio signals), and the behavior of these particles and waves is constrained by the transfer and conservation of energy.

Not every idea in science is crosscutting, which doesn't make it less important—it just means that it's not widely used in every discipline. The theory of evolution by natural selection is a remarkable achievement of science but, by design, applies *only* to biological systems. Biologists can use energy principles to make sense of any living system, but a chemist would not use evolution by natural selection to make sense of atomic bonding, because the principles of evolution simply don't apply to subatomic systems. Likewise, plate tectonics is an enormously useful idea that has provided an essential analytical lens for interpreting geologic phenomena, but, by design, it applies *only* to geologic systems.

---

1 Matter is both a DCI and a crosscutting concept for the same reasons.

Okay, so some big ideas in life science apply only to life science and some big ideas in geoscience only apply to geoscience, but since every system is fundamentally physical, shouldn't every physical science idea be crosscutting? The key factor here is not whether an idea *applies* to phenomena in every discipline but whether it is *useful* for interpreting phenomena in every discipline. For example, *force* is a physical science idea, and living and Earth systems most certainly exert and respond to forces—but conducting a force analysis typically offers little insight into how species change over time or why the atmosphere of the Earth is warming. On the other hand, analyzing energy flow within ecosystems and the atmosphere enhances our understanding of the changes we observe.

Notice that the CCC that appears in the *NGSS* is not just Energy and Matter; it is Energy and Matter: Flows, Cycles, and Conservation. The part after the colon stresses that energy is crosscutting because we can use it to interpret systems based on how it is transferred and transformed (flows) and how energy balances are maintained (cycles), and by carefully tracing energy inputs and outputs (conservation). To fully understand and apply energy as a CCC, it is necessary to understand it as a DCI, because the DCIs about energy provide the tools necessary to fully track flows, cycles, and conservation of energy.

The following sections first discuss key components of energy as a DCI and then the Five Big Ideas about energy that were introduced in Chapter 1 and that open a path to using the disciplinary ideas about energy in a crosscutting way.

Energy as a Disciplinary Core Idea

Within *A Framework for K–12 Science Education* (NRC 2012) and the *NGSS*, energy appears as one of four DCIs in the physical sciences: (1) Matter and Its Interactions, (2) Motion and Stability: Forces and Interactions, (3) Energy, and (4) Waves and Their Applications in Technologies for Information Transfer. Energy is further broken into four component ideas: (1) Definitions of Energy, (2) Conservation of Energy and Energy Transfer, (3) Relationship Between Energy and Forces, and (4) Energy in Chemical Processes and Everyday Life. Each of these component ideas addresses a different aspect of understanding energy as a scientific idea. Like energy, none of these ideas are easy to fully understand—students must build a deeper understanding over time by investigating phenomena in increasingly sophisticated ways.

Definitions of Energy

A good definition of energy is hard to come by. Although many of us develop an intuitive notion that energy helps us do things or that it is associated with action, it is remarkably hard to actually define what energy is. You may have heard that energy is "the ability to do work" or "the capacity to cause a change," but there are problems with each of these popular definitions. Thinking carefully about each of these definitions can illuminate why energy is so hard to define in a way that is meaningful for students.

You may have heard—or even taught students—that work can be calculated by multiplying the force times distance. Easy enough, but then there are all these weird rules for doing work, such as no matter how hard you push or pull on something, as long as it doesn't move (e.g., think of holding a 100-pound dumbbell over your head), then you do no work on the object. Then, there's this requirement that the force has to be in the same direction as the motion to count as work, so when you're carrying a big bag of groceries at a constant speed and height, you're also doing no work on the groceries! Why all the rules? Although work can be *calculated* by multiplying force times distance (as long as they're in the same direction), what work actually measures is the amount by which the energy of a system is changed when a force acts on it. Holding a weight over your head does no work on the weight because when it remains at rest, there is no change in energy in the system containing the dumbbell. Likewise, when you carry a bag of groceries at the same speed and height, the energy of the system containing the grocery bag doesn't change (since neither its speed nor its height above the Earth changes). The upshot of all of this is that defining energy as the ability to do work can end up being pretty confusing for kids since the scientific definition of work can be so nuanced. Even if they're not confused by the concept of work, defining energy as the ability to change the energy of a system is circular and doesn't do much to clarify what energy actually is.

Just as defining energy as the ability to do work is circular, defining it as the capacity to cause a change is so broad that it's barely useful at all. Perhaps worse, it may imply for students that only things that can act on their own volition—living things that can move—can have energy. Given that plenty of students already struggle with this alternative (human-centered) conception, we should think very carefully before offering a definition that may further support this notion.

Richard Feynman said it best when he told his students, "It is important to realize that in physics today, we have no knowledge of what energy *is*" (Feynman, Leighton, and Sands 1989, p. 4-1). While it may seem odd that we use something every day that we cannot define, it's really not all that strange. Consider, for example, how we define a chair. Something that we sit on with four legs and a back, perhaps? Could a chair have three legs? No legs? Does it need a back? What if you are in the woods and you sit on a fortuitously formed tree or rock—does it become a chair? But, somehow, we know a chair when we see it. A chair is useful because of what it does for us, not because it is easy to define. Energy is the same way. Energy helps us track which changes are possible and which are not within a system.

Although we can't define what energy *is*, we can define what it *does*. Energy (whatever it *is*) can be measured in its different manifestations, which we often call forms. For each manifestation of energy, we know how to measure it. When an object moves, we can calculate its energy of motion (i.e., kinetic energy) by multiplying ½ times its mass times its velocity squared ($\frac{1}{2}mv^2$). When we hold an object above the ground, we can calculate the energy stored due to gravity (i.e., gravitational potential energy) by multiplying the mass of the object times the acceleration of gravity times its height above the ground (mgh). So, even though we can't define exactly what energy *is*, we know very well how to recognize it when it changes, and we can calculate these changes very precisely. By tracking changes in energy forms over time, we can identify when energy has been transferred into or out of a system. Throughout K–12, it is much less important to teach students a definition for energy than it is to help them identify and track energy forms in order to use the energy concept to analyze and interpret scenarios.

Definitions of Energy in Elementary School

The *NGSS* recommend that elementary school students *not* be required to learn a definition for energy. While students should learn to associate energy with motion and to recognize that sound, light, heat, and electricity can move energy from one place to another, it is not necessary at this age to introduce definitions for energy or its forms. The *NGSS* focus on motion, sound, light, heat, and electricity because these phenomena are quite natural for students to associate with energy.

Definitions of Energy in Middle School

In middle school, students should begin to formalize the relationship between certain energy forms and the variables that affect them. For example, students should extend their association with motion and energy to understand that the energy of motion is called kinetic energy, that it is proportional to mass, and that it grows with the square of the object's speed. Furthermore, students use the particulate nature of matter to connect heat and kinetic energy by linking an increase in temperature to an increase in the average kinetic energy of particles within an object.

At the middle school level, students begin to learn about potential energy and to associate it with the relative position of objects that interact at a distance via electric, magnetic, or gravitational forces. Students should recognize that potential energy is every bit as real as kinetic energy and that there are many types of potential energy, each associated with a force.

Definitions of Energy in High School

In high school, students begin to understand energy as a fundamentally quantitative concept. By calculating the value of energy forms such as kinetic energy, gravitational potential energy, and thermal energy, students should construct computational models to track energy transfers between systems. Students at the high school level should recognize that all energy is fundamentally the same and that kinetic energy, potential energy, light, heat, and so forth are all manifestations of a single unifying construct called energy.

Students at the high school level should be able to use the atomic theory of matter to connect energy manifestations in both living and nonliving systems. All macroscopic manifestations of energy are ultimately due to the motion of subatomic particles or the energy associated with their interactions. For example, thermal energy is ultimately a manifestation of the random motion of atoms, and chemical energy is associated with the interactions between protons, neutrons, and electrons in a collection of atoms. By understanding energy manifestations at the microscopic scale, students are in a position to consistently use energy as a unifying framework for understanding phenomena across a variety of seemingly disparate phenomena.

Conservation of Energy and Energy Transfer

Energy is only useful as a scientific idea because it is conserved—it is never created or destroyed. A consequence of this fact is that the energy of an isolated system (one that does not interact with its surroundings) will remain the same no matter what happens within it. Thus, the only way to change the energy of a system is through interactions with its surroundings that transfer energy to or from the system.

Students often have trouble accepting the conservation of energy because in our everyday lives there is really no such thing as an isolated system—objects are always interacting with their surroundings. When you get a hot coffee, the fast-moving molecules within it interact via collisions with the slower-moving molecules in the surrounding air, causing molecules in the air to speed up and molecules in the coffee to slow down. Because these collisions decrease the kinetic energy of the molecules in the coffee and increase the kinetic energy of molecules in the air, we say that there is a transfer of energy from the coffee to the air. In other words, your coffee cools down because it transfers energy to the air. If you were able to better isolate your coffee from its surroundings, it wouldn't cool down nearly as fast. An insulated coffee mug is designed to better isolate the coffee system from the surrounding air by minimizing interactions between molecules in the coffee and the air, but even the most expensive and fancy mugs will never fully prevent transfer of energy from the coffee to the air.

When we are young children, we get used to the idea that a hot drink cools down over time because we notice a change in its temperature. But we almost never notice a change in the temperature of the air because there are so many more of the molecules that make up air to speed up than coffee molecules to slow down. While the coffee in your cup may cool down by 100°F, the corresponding increase in temperature of the air in a typical room would be far less than a single degree. As a result, it is hard to think of coffee cooling as an energy transfer; rather, it seems that the energy of the hot coffee simply goes away. But a careful measurement of the surrounding air would reveal that the energy is still there as the coffee cools—it is conserved!

The amazing thing about energy is that by using the principle of conservation, we can get clues that help reveal unseen processes. Even if we never suspected that our coffee heats the air in our house, the principle of conservation of energy requires that the cooling coffee must have some effect on its surroundings, even if this effect is hard to find or to measure. What the law of conservation of energy really tells us is that any time we notice a change in energy of a system, that energy must have been transferred to it or from it. Knowing this, we can look for evidence of changes or processes that may have been previously undetected.

Conservation of Energy and Energy Transfer in Elementary School

Conservation of energy should not be introduced in elementary school. Although it may seem counterintuitive to say that energy should be introduced to students without mentioning its single most important characteristic, it makes sense to delay the introduction of the energy conservation principle because it runs so counter to children's everyday experience. Even if students are able to repeat the phrase "energy is never created or destroyed," a deep understanding of energy conservation really cannot be developed without a rich set of experiences with energy-related phenomena in which energy transfers occur in familiar systems. Examples of such phenomena include collisions between billiard balls, the heating of clothing in sunlight, and the use of a battery to power an electric device. When students explore these phenomena, they should begin to associate the term *energy* (which they have probably already heard and used a lot in their everyday lives) with motion, light, sound, heat, and electricity and to recognize situations in which energy is transferred, such as speeding and slowing, heating by light, and producing sound. These associations can help students recognize that an increase in energy in one object or system is accompanied by a decrease someplace else. For example, when billiard balls collide, one slows down (losing kinetic energy) and the other speeds up (gaining kinetic energy). By associating energy with observable characteristics of objects and systems, students become well positioned to develop a more sophisticated understanding of energy transfer in middle school that ultimately promotes a deeper understanding of energy conservation.

Conservation of Energy and Energy Transfer in Middle School

The *NGSS* stress that middle school instruction should focus on transfers and transformations (or conversions) of energy within and between systems, not on energy conservation. Without using mathematical formulas to calculate numerical values for energy, students should begin to associate the magnitude of various forms (i.e., manifestations) of energy with observable characteristics of objects and systems. In doing so, students can more specifically track how changes in energy within and between systems are related to one another. For example, they should begin to recognize that if an object slows down, there must be some increase in energy somewhere else, such as another object beginning to move or getting hotter. By tracking energy transfers and transformations, students begin to accept that any increase in one form of energy or in one system must be associated with a decrease in another form of energy or another system, and vice versa. Students then begin to understand energy as something that cannot spontaneously appear or disappear. You might think of this idea as the qualitative conservation of energy—it is the characteristic of energy that it can never be created or destroyed. But, rather than focusing on getting students to repeat this statement, middle school energy instruction should be focused on providing students with the evidence that supports it by tracking energy transfers and transformations within familiar phenomena.

Conservation of Energy and Energy Transfer in High School

At the high school level, students are ready to be introduced to the quantitative conservation of energy—that is, the idea that energy is a fundamentally numerical quantity that sets limits on the magnitude of changes that can occur within systems. By calculating numerical values for energy forms such as kinetic energy and energy transferred through processes such as heating, students recognize that energy helps quantify limits on phenomena. For example, energy transfer can help students understand the maximum amount that an exothermic reaction could heat a glass of water, the highest a runaway truck could possibly climb up a hill, or the average minimum amount of vegetation that an herbivore must consume each day to sustain life processes. Students in high school should use quantitative models, in which they calculate specific values for energy forms and transfers, to predict and explain phenomena.

Relationship Between Energy and Forces

Students commonly confuse or conflate the concepts of energy and force, thinking of each as a sort of general capability or impetus for activity. Students commonly think that moving objects have a certain energy or force associated with them, and this energy or force gradually fades away as the object slows to a stop. Children often think of energy or force as a general sort of cause behind why things happen.

Perhaps there is no better example of how students commonly confuse the ideas of energy and force than when Obi-Wan Kenobi explained "The Force" to Luke Skywalker: "The Force is what gives a Jedi his power. It's an energy field created by all living things. It surrounds us, it penetrates us. It binds the galaxy together."[2] Notice that this description not only combines the ideas of force and energy but also implies that energy/force is a product of living things—another common alternative idea about energy! At the risk of second-guessing a Jedi Master, we can do a bit better than Obi-Wan to clarify for students the intimate connection between energy and force.

---

2 Of course, Obi-Wan and Luke Skywalker lived in the fictional world of *Star Wars*. For all we know, Jedi Master Obi-Wan Kenobi may very well have been speaking correctly about their universe—just not the one we live in today.

The concepts of force and energy are connected in two primary ways: potential energy and energy transfer.

Forces and Potential Energy

Every type of potential energy is associated with a force. For example, gravitational potential energy is associated with the force of gravity that exists between objects. If you stand on top of a table, you have gravitational potential energy relative to the ground. How can you tell? Because if you step off of the table, you will gain kinetic energy as you fall. The higher up the tabletop is, the faster you will be going when you hit the ground—that is, the more kinetic energy you will have when you land. Any time you are above the ground, you have gravitational potential energy relative to the ground because if you fall to the ground, you will gain kinetic energy as the force of gravity acts on you. If there were no force of gravity, you would not fall toward the ground. Two things are important to notice about gravitational potential energy: (1) gravitational potential energy exists because of the force of gravity between you and the Earth, and (2) the way to change your gravitational potential energy is to get farther from or closer to the Earth.

Every form of potential energy shares two features in common: (1) it exists because of the force acting between two objects, and (2) its magnitude depends on the arrangement of the objects relative to each other. Elastic potential energy arises from the electromagnetic force between the charged particles in atoms and molecules that make up elastic materials, and the amount of elastic potential energy depends on how stretched or compressed the material is—which affects how far apart the atoms and molecules are from each other inside of the material. Likewise, chemical energy is a form of potential energy and arises because of the electromagnetic forces between charged particles in atoms and molecules. When the arrangement of atoms changes in a chemical reaction, so does the amount of chemical potential energy associated with that arrangement. Potential energy, regardless of its type, exists because of the force between interacting objects and changes when the arrangement of those objects is altered.

Forces and Energy Transfer

Every force is mediated by something called a field. You have seen evidence of a field if you have ever seen the pattern formed by iron filings in the vicinity of a magnet, as shown in Figure 2.1 (p. 28). Each tiny iron filing is affected differently by the magnet depending on its location relative to the magnet. The iron filings form the pattern that they do because there is a magnetic field that permeates the space around it, and this magnetic field causes the iron filings to point in specific directions. If you bring two magnets close to each other, they will attract or repel based on how their magnetic fields interact with each other. The magnets can push or pull on each other even though they are not touching because a force is conveyed by the magnetic field around the magnet.

Figure 2.1. Iron filings in a magnetic field

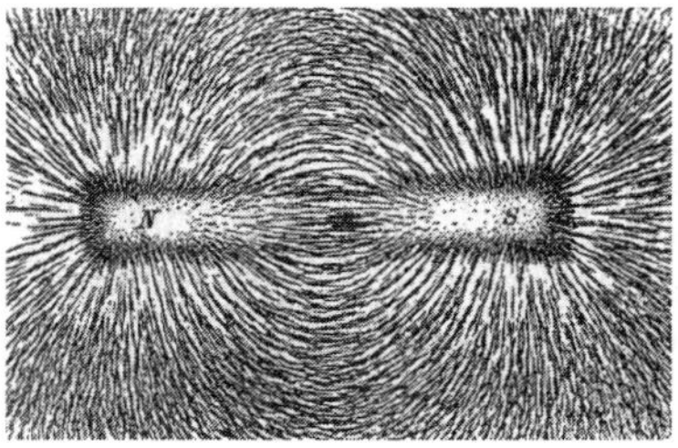

Source: Berndt Meyer. *https://upload.
wikimedia.org/wikipedia/commons/5/57/
Magnet0873.png*

Likewise, electrically charged particles like protons and electrons have electric fields associated with them. These electric fields permeate the space around them and mediate the repulsive force between like charges and the attractive force between unlike charges. Every interaction between objects is mediated by fields, thus every energy transfer is accomplished via fields. When two objects collide, the force between them is really due to the electric and magnetic fields (which are actually part of the same electromagnetic field) that exist between the charged particles that make up the atoms and molecules within each object. Remember the example given earlier of billiard balls colliding such that one slows down and the other speeds up; this transfer is conveyed via changes in the electromagnetic field between the atoms in the colliding objects.

Just as electric and magnetic forces are mediated by electric and magnetic fields, the gravitational force is mediated by a gravitational field that acts between all objects. The Earth pulls all of us toward it (and we pull upward on the Earth) because of the gravitational field that exists between the Earth and our bodies. The gravitational field is actually quite similar to the electric field, except that it is *much* weaker (it takes a lot of mass—about a planet's worth—to create a substantial gravitational force) and it is *always* attractive (i.e., gravity only pulls objects toward each other—it never repels them).

Sometimes energy can be transferred from a system via fields without any other object immediately receiving the energy. When a charged particle moves back and forth, this changes the electric field in the space around it, and this changing electric field generates a magnetic field. What's really interesting is that this magnetic field returns the favor—it generates an electric field as it changes. This process of electric and magnetic fields co-generating each other as they change results in a self-propagating disturbance through space that we call an electromagnetic wave, or light. If these traveling fields encounter charged particles, they interact with them and affect their motion. When this change in motion occurs, there is a corresponding transfer of energy to the particles the electromagnetic wave encountered.

Thus, electromagnetic waves can transfer energy through space between two systems, even when they are not physically in contact with each other.

Because the charged particles that make up atoms are always in motion—as atoms and molecules move and vibrate in matter—these vibrating charges are always emitting electromagnetic waves in a process called electromagnetic radiation. This electromagnetic radiation depends on the temperature of objects, since temperature indicates how quickly atoms and molecules move in matter. Thus, many people refer to this as thermal radiation, since it is associated with the random movements of atoms and molecules in matter. Since atoms and molecules are always in motion, all objects are constantly emitting thermal radiation. Thermal imaging cameras and some animals, such as pit vipers, rely on this fact to detect warm objects and living organisms. This detection is possible because all objects with a temperature (in other words, all objects!) transfer energy to their surroundings via electromagnetic radiation.

Fields and Energy

The notion of fields helps connect the concepts of force and energy. Fields can store energy in potential energy or transfer energy between systems. Because every potential energy is associated with a force, which is conveyed by a field, we can think of potential energy as being stored in the fields between the objects that have forces between them. Gravitational energy is stored by the gravitational field between interacting objects. Chemical energy is stored in the electric field between charged particles in atoms and molecules.

Fields can also transfer energy through their associated forces. In an electric circuit, an electric field is generated in a conducting wire that pushes charges along the wire. In a microwave oven, electromagnetic waves (which are oscillating electric and magnetic fields) transfer energy to water molecules as these waves cause water molecules in food to vibrate and turn. As the more energetic water molecules collide with the molecules around them, the temperature of the food increases.

Fields are a truly valuable idea in science and serve to connect the concepts of force and energy, but they are not an appropriate concept for all learners. Although the goal in K–12 education is to help students use the idea of fields to connect force and energy, these ideas must be built over time.

Relationship Between Energy and Forces in Elementary School

In elementary school, students are just starting to become familiar with the idea of a force and to understand energy as a scientific idea. Students at this age should focus on recognizing that objects can push or pull on each other (e.g., during collisions) and, in the process, transfer energy from one to the other. Later in elementary school, students should focus on the idea that some forces (such as that of magnets) can be exerted even though objects are not in contact with each other, and this interaction can change the kinetic energy of the interacting objects. Although they do not explicitly connect the scientific concepts of energy and force (i.e., by relating each to fields) at this age, students should notice that energy transfers occur between objects when the objects push or pull on each other.

Relationship Between Energy and Forces in Middle School

In middle school, students build on their understanding of the role of energy and forces in phenomena that transfer energy and connect forces to the notion of potential energy. Students should recognize that when two objects interact at a distance through magnetic, electric, or gravitational forces, there is potential energy stored within the system. In addition to associating potential energy with forces that act between objects, students should recognize that energy can be transferred between systems by forces that act at a distance. Further, they should be able to identify energy transfers that occur as a result of electromagnetic radiation, but at this point they don't need to associate this radiation with the notion of a propagating field oscillation.

Relationship Between Energy and Forces in High School

In high school, students should begin to use the concept of fields to associate force and energy. They should relate potential energies to the associated force field and should recognize that propagating field oscillations can transmit energy through space and between systems. By creating and using models of fields (e.g., electric field lines drawn through space), students should explain the role of fields in storing different amounts of energy or transferring energy between systems.

Energy in Chemical Processes and Everyday Life

Students of all ages encounter the energy concept in their everyday lives, perhaps most commonly in the context of food or fuel. Many students think of food and fuel as containing energy, and they often believe that when food is digested or fuel is burned, the energy contained within the food or fuel is used up. This idea is not surprising: We can't burn gasoline twice, batteries don't spontaneously recharge themselves, and, of course, we can't just keep recycling the same plate of pasta to sustain our life functions.

Older students often think that energy in food and fuel is stored in chemical bonds, and that this energy is released when the bonds are broken—you may have heard or even taught this idea. Although energy can indeed be released in chemical reactions, and bonds are indeed broken in these chemical reactions, it is not the case that chemical bonds somehow contain energy. In fact, atoms form bonds in the first place because the bonded arrangement is associated with a *lower potential energy* (due to the electric force field between the charged particles in atoms and molecules) than the unbounded arrangement. Likewise, a chemical reaction makes it possible for atoms to rearrange in such a way that they have an even lower potential energy. Just like a ball that speeds up as its gravitational potential energy decreases, when chemical potential energy decreases, this "missing" energy must be manifested in some other way, such as faster-moving molecules. We detect this increase in faster-moving molecules as a change in temperature.

When we use food or fuel, molecules in the food or fuel react with oxygen to form a new set of molecules with an arrangement that has a lower potential energy than the original arrangement. If you drive a gasoline-powered car, fuel reacts with oxygen in the piston of your car's engine, lowering the chemical potential energy and resulting in molecules that move much faster and therefore expand rapidly. This rapid expansion happens because the molecules are moving much faster and have more kinetic energy; these more energetic molecules collide with the piston and exert a large force on it, which pushes the piston up. Through a clever system of alternating pistons, gears, and axles, this upward motion of the piston can be translated into forward motion of the car.

When food or fuel reacts with oxygen, this reaction produces carbon dioxide and water, which are molecules with a very low potential energy arrangement. Because these arrangements have such low energy, they are no longer useful as fuel. Thus, it is tempting to think that fuel stores energy and water or carbon dioxide does not. But water can react with some substances and release a great deal of energy. For example, drop pure sodium into water and you will see quite a release of energy! But nobody is suggesting that this reaction should be used to power cars, because sodium doesn't exist in nature in its pure form—it would take a lot more energy to purify the sodium than we get out when it reacts with water. Fossil fuels, on the other hand, are found in nature and contain molecules that react readily with another substance found in nature—oxygen in the atmosphere. All we need to do is give it enough heat to get the reaction going, and the molecules contained within the fossil fuel (e.g., methane, octane) will continue reacting with oxygen to form molecules with a lower potential energy arrangement and, thus, more kinetic energy. These faster-moving molecules keep the reaction going, can be used to make changes in the surroundings, and tend to spread out. If they are trapped, such as in a piston or a steam engine, they can be very useful for driving machines; if they are not trapped, their tendency to spread out makes them very difficult to harness and do something useful—so difficult, in fact, that this energy is essentially "gone" if the molecules are not trapped. Although this energy is not really useful, it still very much exists.

Students commonly think that the energy in food or fuel is gone and no longer exists afterward, but we know that a careful tracking of the energy conversions and transfers that take place when food or fuel is used will reveal that there is the same amount of energy as there was before the reaction, but it now exists largely in the thermal energy that is associated with faster-moving molecules. In everyday life, *conserving energy* means to use less fuel. But in science, *conservation of energy* means that energy is never created or destroyed—or, more precisely, that the total energy within a closed system will not change unless there is a transfer into or out of it.

The Earth is not a closed system. Energy is transferred to the Earth via sunlight and re-radiated into space as infrared radiation. A small fraction of the energy transferred to the Earth via sunlight is stored by plants as chemical energy when they produce food via photosynthesis. Over time, this energy can be transferred to other organisms or stored in Earth materials as fossil fuels. Thus, sunlight is ultimately the major source of energy for powering our lives. But other sources are important, too. Uranium in the Earth can be mined and used to power nuclear reactors, using the energy stored as potential energy that arises from the nuclear force between protons and neutrons in the uranium nucleus. The thermal energy associated with hot materials deep within the Earth can also be used to power electric generators. There is energy all around us, but not all of it is easy to access for powering the various devices that we use every day of our lives. The term *energy resources* refers to natural resources that are exploited relatively easily in order to transfer energy from the natural resource to the devices that we wish to power. Although energy itself may be conserved, many natural energy resources—most notably fossil fuels—most certainly are not. It would be much more accurate to urge people to help conserve energy *resources* rather than help conserve energy.

Energy in Chemical Processes and Everyday Life in Elementary School

At the elementary level, students should focus on identifying energy inputs and outputs within living and designed systems. By recognizing that portable electronic devices such as flashlights or cell phones need batteries, animals need food, and plants need light, students begin to build an understanding that such systems and organisms need an energy input. Students should also begin to recognize that energy is transferred to the environment via heat in many phenomena, such as an animal moving or a computer operating. At this age, students should not focus on the mechanism by which energy conversions take place within organisms and devices. Although they should recognize that plants need sunlight, a discussion of the process of photosynthesis should wait until middle school.

Energy in Chemical Processes and Everyday Life in Middle School

In middle school, students learn about the process of photosynthesis as a chemical reaction in which energy from sunlight is converted into chemical energy stored in food made by plants. This energy can be released by the processes of digestion or burning. Additionally, students should build on their understanding that organisms and devices transfer energy to their surroundings through heat and begin to recognize that some processes result in more heat transfer than others. By recognizing that some processes transfer more heat than others, this sets the stage for learning about efficiency in later grades and comparing the value of energy resources.

Energy in Chemical Processes and Everyday Life in High School

At the high school level, students should explore the processes by which energy is converted within living and designed systems and to evaluate the efficiency by which these conversions occur. By evaluating processes by which organisms and devices use energy and considering factors such as efficiency, availability, and waste products, students should be able to compare different energy resources and consider the benefits and drawbacks of their use.

The Five Big Ideas About Energy

In our discussion of energy as a DCI, you probably noticed that a few ideas kept coming up. Energy is a CCC as well as a DCI because these few ideas are commonly used to make sense of phenomena across all scientific disciplines. When used in practice, they can look quite different, but these ideas form a basis for knowing how to use the energy concept to understand phenomena from all disciplines. The Five Big Ideas introduced in Chapter 1 are repeated here and then discussed in detail in this section.

- **Big Idea 1.** All energy is fundamentally the same, and it can be manifested in different phenomena that are often referred to as different "forms" or "types."
- **Big Idea 2.** Energy can be transformed/converted from one form/type to another.
- **Big Idea 3.** Energy can be transferred between systems and objects.
- **Big Idea 4.** Energy is conserved. It is never created or destroyed, only transformed/converted or transferred.
- **Big Idea 5.** Energy is dissipated in all macroscopic (involving more than just a few particles) processes.

Big Idea 1. Energy Can Be Manifested in Different Forms.

This Big Idea says that not only are various forms of energy (e.g., kinetic energy, thermal energy, gravitational potential energy, and nuclear energy) related to one another—they are exactly the same thing! Said differently, this Big Idea means that energy is energy is energy. Potential energy is not something that is waiting to be activated in order to become energy—it already *is* energy, just as much as kinetic energy or thermal energy is. This Big Idea says that energy is a fundamental underlying property of a system and that the forms of energy that we see are simply manifestations of this single underlying property.

Big Idea 2. Energy Can Be Converted From One Form to Another.

This Big Idea follows from the first one. If the forms of energy are all manifestations of the same thing, then they must be able to change into one another. Big Idea 2 says that there are no forbidden conversions. Any form of energy can become any other form.

Big Idea 3. Energy Can Be Transferred Between Systems and Objects.

Energy can go places. The energy of one system or object is entirely interchangeable with the energy of another system or object. Living and nonliving systems can exchange energy just as well as colliding billiard balls can. Like Big Idea 2, which says that all forms of energy are fundamentally the same, this Big Idea says that the energy of all systems is fundamentally the same and can move between them.

Big Idea 4. Energy Is Conserved.

This is by far the most important of the Five Big Ideas. You can think of this as "energy whack-a-mole." Anytime we see a decrease in one form of energy or in the energy of a system, it must pop up somewhere else. Likewise, an increase in a form or system must occur in combination with a decrease somewhere else. The amount of decrease must be exactly the same as the amount of the increase, and vice versa. Knowing this idea, we can set limits on the behavior of systems by knowing which things are possible and which are not.

Big Idea 5. Energy Is Dissipated in All Macroscopic Processes.

This idea is why it can be so hard to believe Big Idea 4. This idea explains why so many systems "leak" energy into their surroundings and why energy seems to be used up or go away over time, and that energy dissipation happens because nature tends toward an even, equilibrated state. Molecules are always colliding with one another, and these collisions happen in all directions and across system boundaries. If something is hot, there are a lot of fast-moving molecules in one place, and these molecules will heat up their surroundings as they collide across system boundaries and transfer their energy. This energy transfer will continue until faster-moving molecules are evenly spread out within the object and its surroundings. Light and sound also transfer energy. Like ripples in a pond, sound and light waves spread out in all directions, reflect off boundaries, and carry energy with them as they propagate.

As energy spreads out, it can do so quickly or slowly. If something is a lot hotter than its surroundings, it will cool down really quickly (e.g., a freshly baked pie cools more quickly in the freezer than on the counter—but won't necessarily taste better!). Similarly, if something impedes the transfer of energy, it takes longer to spread out (e.g., an insulating coffee mug is designed to separate hot coffee molecules from their surroundings, slowing down its rate of cooling). Whether it happens quickly or slowly, the spreading out of energy can often be difficult to notice.

Pour a bottle of water into a bucket, and you will see the water level of the bucket rise. Pour it into a bathtub, you might see a difference. Pour it into a backyard pool, good luck. Lake Superior, never. The more room the water has to spread out, the harder it is to detect that you added the water in the bottle. Big Idea 5 reminds us that energy spreads out within systems and readily transfers between systems, and that this spreading can make energy maddeningly difficult to detect.

Energy as a Crosscutting Concept

The Five Big Ideas are directly applicable in every scientific discipline and form the foundation of using energy as a CCC. Of course, the *NGSS* emphasize energy flows, cycles, and conservation as major ideas that make energy crosscutting. The Five Big Ideas about energy form the foundation by which students become capable of analyzing and interpreting energy flows, cycles, and conservation in a variety of phenomena across disciplines.

Using the Five Big Ideas in Elementary School

In the Energy and Matter: Flows, Cycles, and Conservation CCC, energy is not foregrounded in elementary school—the major focus is on matter. But students need to learn about energy in elementary school, and it is important for energy instruction to set the stage for using the concept both within the physical science discipline and across disciplinary contexts. Elementary teaching should focus on recognizing energy manifestations such as heat, light, sound, electricity, and motion and that they are related to each other (Big Idea 1). For example, motion can produce heat, headphones need electricity to produce sound, and light can heat up food and clothing. Connecting these phenomena forms a powerful foundation for students to recognize that they are all manifestations of the same underlying concept.

By exploring interacting objects and systems, students should observe evidence of energy transfer (Big Idea 3). A hot object can heat up a cooler one, and, in the process, the hotter object cools down. A fast-moving object can speed up a slower one (e.g., in a collision), and, in the process, the faster object must slow down. Students at the upper elementary level can think about energy transfer by asking questions such as "Where does the energy come from? Where does the energy go?" to analyze phenomena such as collisions between marbles or the heating of water by putting a warm object in it (Crissman et al. 2015).

Using the Five Big Ideas in Middle School

In middle school, students should begin to explicitly recognize that energy forms such as light, heat, kinetic energy, and gravitational potential energy are fundamentally the same thing called energy (Big Idea 1), and they should be able to identify evidence for when these energy forms are increasing or decreasing. They should use this ability to recognize that if one form of energy increases, this increase must be coupled with a decrease in at least one other form (Big Idea 2) or a transfer from outside of the system (Big Idea 3). Although recognizing these coupled increases and decreases builds evidence for the conservation of energy, the focus is not yet on conservation.

Students should identify evidence of energy transfers and manifestations of energy within familiar phenomena such as growing plants, evaporating water, or swinging pendulums. In the process, students should learn to recognize that cyclical processes, living systems, and powered devices will cease to function without a consistent energy input, and they should connect this cessation with energy transfers out of the system. Recognizing that most everyday systems stop functioning without a consistent energy input sets the stage for understanding energy dissipation in later grades.

Students in the middle grades should focus on recognizing energy forms (Big Idea 1), conversions (Big Idea 2), and transfer (Big Idea 3). This focus builds an evidence base for understanding energy conservation (Big Idea 4) and dissipation (Big Idea 5), but middle school students are not yet expected to use the principles of energy conservation or dissipation to explain phenomena.

Using the Five Big Ideas in High School

High school students should both recognize energy forms in a wide range of phenomena and use numerical formulas to calculate their values (Big Idea 1). In doing so, they quantitatively track energy conversions and transfers (Big Ideas 2 and 3) and model the quantitative conservation of energy (Big Idea 4).

By studying physical systems that are very nearly closed (i.e., no transfer into or out of the system), such as a massive pendulum swinging with a small amplitude, students can test the predictions of computational models that assume quantitative energy conservation and also recognize that the discrepancy between the model predictions and real-world measurements continues to grow over time until the systems stops moving, which provides evidence for energy dissipation (Big Idea 5) and enables students to calculate a numerical value for how much energy has been dissipated.

By the end of high school, students should be able to use quantitative models that track energy transfers into and out of living and nonliving systems and relate changes within energy forms within these systems to one another. By creating quantitative models of energy changes and transfers, students use the same analytical techniques to interpret and explain a wide range of phenomena.

Summary

Energy is an idea that comes out of the physical sciences, but it is critical to analyzing and interpreting phenomena from all scientific disciplines. Too often, however, instruction in each discipline fails to give students the tools to connect their ideas about energy from one discipline to the next. The *NGSS* stress the importance of teaching energy both as a DCI in the physical sciences and as a concept that cuts across disciplines. Teaching energy as a CCC does not mean that biology teachers should start teaching physics lessons in their classroom or that chemistry teachers should maintain a class aquarium. Rather, teaching energy as a CCC means that teachers in all grades, whether discipline-specific or not, should present energy ideas in a way that is consistent across disciplines and promotes, rather than impedes, students' ability to use what they know about energy as a physical science idea to make sense of phenomena in the life sciences and vice versa.

By intentionally designing energy instruction that helps students understand and apply the Five Big Ideas about energy, you emphasize ideas that form an analytical lens to represent how energy can be used consistently to interpret a wide range of phenomena in both scientific and everyday settings. In the next chapter, we use a popular science activity to illustrate the difference between teaching energy as a discipline-specific idea and presenting energy as a crosscutting concept.

References

Chabalengula, V. M., M. Sanders, and F. Mumba. 2012. Diagnosing students' understanding of energy and its related concepts in biological context. *International Journal of Science and Mathematics Education* 10 (2): 241–266. *http://doi.org/10.1007/s10763-011-9291-2.*

Crissman, S., S. Lacy, J. C. Nordine, and R. Tobin. 2015. Looking through the energy lens. *Science and Children* 52 (6): 26–31.

Driver, R., A. Squires, P. Rushworth, and V. Wood-Robinson. 1994. *Making sense of secondary science: Research into children's ideas.* New York: Routledge.

Duit, R. 2014. Teaching and learning the physics energy concept. In *Teaching and learning of energy in K–12 education,* eds. R. F. Chen, A. Eisenkraft, D. Fortus, J. S. Krajcik, K. Neumann, J. C. Nordine, and A. Scheff, 67–85. New York: Springer.

Feynman, R. P., R. B. Leighton, and M. L. Sands. 1989. *The Feynman lectures on physics.* Vol. 1. Redwood City, CA: Addison-Wesley.

Liu, X., and A. McKeough. 2005. Developmental growth in students' concept of energy: Analysis of selected items from the TIMSS database. *Journal of Research in Science Teaching* 42 (5): 493–517.

Miller, E. 2014. Exploring crosscutting concepts. In *Introducing teachers and administrators to the NGSS: A professional development facilitator's guide,* 79–82. Arlington, VA: NSTA Press.

National Research Council (NRC). 2012. *A framework for K–12 science education: Practices, crosscutting concepts, and core ideas.* Washington, DC: National Academies Press.

Neumann, K., T. Viering, W. J. Boone, and H. E. Fischer. 2013. Towards a learning progression of energy. *Journal of Research in Science Teaching* 50 (2): 162–188. *http://doi.org/10.1002/tea.21061.*

NGSS Lead States. 2013. *Next Generation Science Standards: For states, by states.* Washington, DC: National Academies Press. *www.nextgenscience.org/next-generation-science-standards.*

Nordine, J., J. Krajcik, and D. Fortus. 2011. Transforming energy instruction in middle school to support integrated understanding and future learning. *Science Education* 95 (4): 670–699. *http://doi.org/10.1002/sce.20423.*

Smith, J. P., A. diSessa, and J. Roschelle. 1993. Misconceptions reconceived: A constructivist analysis of knowledge in pieces. *Journal of the Learning Sciences* 3 (2): 115–163.

Solomon, J. 1983. Messy, contradictory, and obstinately persistent: A study of children's out-of-school ideas about energy. *School Science Review* 65 (231): 225–229.

Watts, M. 1983. Some alternative views of energy. *Physics Education* 18: 213–217.

CHAPTER 3

TEACHING ABOUT ENERGY AS A CROSSCUTTING CONCEPT

ARTHUR EISENKRAFT

Woe the teacher who asks students, "Why do we eat?" and expects a simple science answer such as "to keep warm and active." We eat because we are hungry. We eat before we are hungry. We eat because food tastes good. We eat because we want to eat—especially chocolate, potato chips, and ice cream.

We do know implicitly that we eat to be healthy and that not eating would be difficult and painful. We also know that food has calories (whatever they are) and that it's important not to eat too many calories or we will become overweight.

We can investigate food and calories in a science class with the popular cheese puff lab experiment.[1] A cheese puff is burned underneath a container of water; as the cheese puff burns, the water heats up. The cheese puff is a fuel that can keep a flame going and raise the temperature of water. When we *eat* cheese puffs, our body uses that same fuel to keep our body temperature at 98.6°F even when we sit in rooms that are 68°F.

As students mature, this same experiment can be repeated at different levels of sophistication. In elementary school, students can explore how burning (digesting) food provides (transfers) energy for their bodies and mind. In the middle grades, students can compare the temperature changes of water that occur when burning different foods. In high school, students can determine the energy content of a cheese puff from an analysis of the combustion reaction.

This same investigation can be linked to biology (digestion), chemistry (combustion), and physics (conservation laws). The lab can be looked at from an energy perspective and provide insights into energy transfer, transformation, conservation, and dissipation. As such, we focus on energy as both a disciplinary core idea (DCI) and a crosscutting concept (CCC).

Many science teachers recognize energy as an idea that is core to their discipline. When teaching physical science, they guide students to the idea that a fast-moving object has

Acknowledgment
I would like to acknowledge contributions to early drafts of this chapter by Fernando Cleves, Haven Daniels, Michael Harris, Angela Palo, and Dean Martin of the Boston Public Schools.

1 Although more people may be familiar with peanuts than with cheese puffs, the cheese puff lab has replaced the "peanut lab" in some classrooms because of the risk of peanut allergies.

more energy than a slow-moving object, a bright light emits more energy than a dim light, a hot object has more energy than a cool object, and a loud sound emits more energy than a soft sound. When teaching life science, teachers help students understand that living things require an energy source, which, for plants, can be the Sun and, for animals, can be plants. As students move from kindergarten to grade 12, these ideas become more sophisticated and more connected to mathematics.

It is because energy is critical to every discipline that it is also a CCC; it is a concept that transcends disciplinary boundaries. Energy is a key consideration for making sense of phenomena in physical sciences (chemistry and physics), Earth and space sciences, and life sciences. Energy is a theme that we keep returning to regardless of the science we study. For example, one could attend a science meeting and at the conclusion of almost any paper, you could ask the speaker, "What are the energy considerations of your work?" and it will probably be an appropriate question.

The fact that energy can be transferred between systems is an idea used in all branches of science, which is an example of energy as a CCC. In contrast, learning the mathematical definition of kinetic energy, calculating the binding energy of an electron, or learning how a catalyst can lower the activation energy of a reaction are examples of energy as important DCIs.

Unfortunately, the overarching importance of energy as a CCC is often overlooked in our instruction. The emphasis on energy as a CCC in *A Framework for K–12 Science Education* (NRC 2012) and the *Next Generation Science Standards* (*NGSS;* NGSS Lead States 2013) is an attempt to change this instructional practice and provide students with explicit references to how energy is one concept across all sciences, regardless of whether we are teaching a unit in physical science, Earth and space science, or life science.

This chapter describes the popular cheese puff lab, which is often treated as either a physical science investigation or a life science investigation, and illustrates how it can be recast to emphasize energy. This illustration will serve as an example of how treating energy as a CCC can be accomplished in other lab investigations. Later in this chapter, lesson plans for the cheese puff lab using the 7E instructional model (Eisenkraft 2003) are presented.[2]

As educators, we have commonly taught energy as a DCI. This chapter illustrates how typical instruction can be recast to emphasize energy as the CCC that it is.

The Traditional Cheese Puff Lab

The cheese puff lab is traditionally presented by showing students a food label's nutrition facts and asking, "How do they get the number for calories to put on the food label?" The research question for the students is typically something such as, "Is the energy content

---

2 Although the 7E instructional model is used here, many other instructional models can be used to emphasize energy as a crosscutting idea.

(calories) on the food label correct?" This question can be modified to be made appropriate for students at different grade levels. For example, a first-grade class can be shown that two different food labels have different values for the number of calories and can be asked, "How do the food scientists know what number to put on the label?"

The students use an apparatus that has a place for a cheese puff to burn and a beaker of water above the cheese puff (Figure 3.1).

Figure 3.1. Apparatus for burning a cheese puff

As the cheese puff burns, the water's temperature rises. By measuring the temperature increase of the water, the energy gain of the water can be calculated. The energy gain of the water is then typically assumed to be equal to the energy loss of the burning cheese puff. In this way, the energy content of the cheese puff can be found.

Different versions of this investigation are done in elementary school, in middle school, and in chemistry, physics, and biology classes in high school. The investigation can have a more sophisticated apparatus (e.g., a calorimeter replaces the beaker), more calculations, and more attention to energy losses to the environment, depending on the grade-level learning expectations.

Safety Notes

1. Use sanitized indirectly vented chemical-splash goggles and aprons during pre-lab setup, during the activity, and during post-lab cleanup.

2. Use caution in working with an active flame. It can burn skin and clothing, so keep hands and clothing away from the flame.

3. Make sure the site is free of flammable liquids.

4. The ring stand, beaker, and heated water will have elevated temperatures and heat levels when heated. Wait for them to cool down before handling.

5. Remind students not to eat any food used in a lab activity.

6. Wash hands with soap and water after completing the lab.

The lesson can include, but often may not, many of the components of the *Framework*. In the science and engineering practices dimension, the lesson can include the following practices:

- Asking questions (for science) and defining problems (for engineering)
- Designing and using models
- Planning and carrying out investigations
- Analyzing and interpreting data
- Using mathematics and computational thinking
- Constructing explanations and designing solutions
- Engaging in argument from evidence
- Obtaining, evaluating, and communicating information

In the CCCs dimension, the lesson can include the following concepts:

- Cause and effect: Mechanism and explanation
- Systems and system models
- Energy and matter: Flows, cycles, and conservation
- Stability and change

In the DCIs dimension, the lesson can include the following core ideas:

- Physical sciences
  - Energy (PS3), including definitions of energy (PS3.A), conservation of energy and energy transfer (PS3.B), relationship between energy and forces (PS3.C), and energy in chemical processes and everyday life (PS3.D)
- Life sciences
  - From molecules to organisms: Structures and processes (LS1), including structure and function (LS1.A) and organization for matter and energy flow in organisms (LS1.C)
  - Ecosystems: Interactions, energy, and dynamics (LS2), including cycles of matter and energy transfer in ecosystems (LS2.B)
- Engineering, technology, and applications of science
  - Engineering design (ETS1), including defining and delimiting an engineering problem (ETS1.A), developing possible solutions (ETS1.B), and optimizing the design solution (ETS1.C)

- Links among engineering, technology, science, and society (ETS2), including interdependence of science, engineering, and technology (ETS2.A) and influence of engineering, technology, and science on society and the natural world (ETS2.B)

That all of these science and engineering practices, CCCs, and DCIs can be included in the cheese puff lab does not imply that they all should be included. This investigation can, and should, be focused on a smaller set of ideas than the lists presented above. The teacher (with *NGSS* for guidance) has to choose which of these practices, CCCs, and DCIs should be central to this investigation and other investigations throughout the year, while ensuring that all are treated during each year of science study. The next section illustrates how the traditional cheese puff lab can be adapted to emphasize the crosscutting nature of the energy concept.

The Revised Cheese Puff Lab With the Crosscutting Aspects of Energy as a Central Focus

The *Framework* calls for integrating the three dimensions of science learning (i.e., science and engineering practices, CCCs, and DCIs) in our curriculum. Because CCCs are a part of all science content and scientists think about all science content with respect to these concepts, it is important that students be consistently reminded of these concepts throughout each grade and within a wide variety of science lessons. This will help them connect what they have learned in third grade with what they learn in eighth grade and what they learn in biology with what they learn in chemistry. Having the view that all science is built on the same concepts will help students better organize the DCIs into a larger structure that transcends any one discipline. The *Framework* mirrors earlier standards documents in their call for students to recognize the unifying concepts or common themes of science while learning the core ideas of science (AAAS 1993; NRC 1996).

CCCs such as energy, systems, and stability and change should be emphasized across disciplines and across grades and repeated many times throughout the year. The *Framework* and the *NGSS* stress the need to emphasize these concepts in a single lesson and, then, in many lessons. Teachers commonly mention these CCCs (or assume that they have mentioned them) in their instruction, but these concepts have typically been in the background. Although we probably all agree that there is a need to focus on CCCs, it can be difficult to put this emphasis into practice because it requires teaching about energy in new ways. This chapter focuses on the cheese puff lab as an example of an opportunity to emphasize the crosscutting nature of energy in science instruction. Revising the traditional approach to this lab can both clarify the meaning of energy and serve as an example of how to capitalize on opportunities to focus on the crosscutting nature of energy throughout the year.

Many teachers may think that energy is already in the foreground of the cheese puff lab. In the traditional cheese puff lab, the energy gain of the water is calculated after measuring the change in temperature of the water. The energy content of the cheese puff is then determined by *assuming* that the energy loss of the burning cheese puff is equal to the energy gain of the water. The lesson may have begun by asking students how the value on the calorie label on the package of cheese puffs is determined. The lesson may end with students comparing their value from the in-class laboratory investigation with the value as given on the food label and calculating percent error. Students may then be asked to provide reasons for their value being different from the food label value.

This traditional cheese puff lab *assumes* the conservation of energy but does not emphasize it as a major concept or CCC in science. The energy concept is in the background, while the calculation of the energy content of a food product is in the foreground. Unfortunately, students walk away from the lab with the conclusion that the calorie count of the cheese puff as determined from their experiment is less than the calorie count on the food label. It would be preferable for them to walk away from the lab concluding that the energy transferred to the water is less than the energy transferred from the burning cheese puff, *and* since energy must be conserved, this difference in energy must be accounted for in other ways.

The cheese puff lab can emphasize the Five Big Ideas about energy (i.e., forms, transformation, transfer, conservation, dissipation). The forms of energy include chemical energy in the cheese puff and thermal energy in the water. There is a transformation of energy from chemical energy in the cheese puff–air system to the thermal energy and light, as well as an accompanying transfer of energy from the cheese puff–air system to the water-container system. Throughout the process, the total energy of all involved systems is conserved. Any difference in energy between the caloric value of the cheese puff and the thermal energy of the water in energy is due to the transfer and dissipation of energy in the surroundings and the apparatus and to the transfer of energy associated with light leaving the flame.

In the revised cheese puff lab, the shift in the lesson is to ask students whether energy is conserved when a cheese puff is burned to raise the temperature of water. This puts the CCC of energy conservation front and center. The students then design their investigation in a manner that may be nearly identical to the traditional lab. At the high school level, the students measure the increase in temperature of the water and calculate the energy gain of the water. They use the value of the energy content of a cheese puff from the food package as a given. They then ask if the energy loss of the burning cheese puff is equal to the energy gain of the water. It is not! The students then have to explore what could have gone wrong. Different claims can be generated from the class discussion: Is conservation of energy wrong? Is the food label incorrect? Were errors made in the calculation? Are there energy transfers that have not been taken into account?

The revised cheese puff lab frames the conservation of energy as the research question of the lesson. It is the important concept, and the energy content of the cheese puff is used to verify whether energy is conserved. In contrast to the traditional lab, which has a focus on calculation and determining if students can replicate in the lab the value on the food label, the revised lab has energy as its focus and asks students if energy is conserved (see Table 3.1). Placing an investigation of the conservation of energy in the foreground of the investigation is one way to treat energy as a CCC or unifying concept of all science.

Table 3.1.

COMPARISON OF THE TRADITIONAL CHEESE PUFF LAB WITH THE REVISED LAB

Traditional lab	Revised lab
Burn cheese puff and heat up water.	
Research question: Is the energy content (calories) of the cheese puff on the food label correct?	Research question: Is energy conserved? Is the energy loss of a burning cheese puff equal to the energy gain of water?
Measure the temperature gain of the water. Calculate the energy gain of the water.	
Assume that energy is conserved. (Often this assumption is not stated at all.)	Calculate the energy loss of the cheese puff using the food label information.
Determine the energy content of the cheese puff from the assumption that energy is conserved. (Again, this assumption is often not stated explicitly. Students are just told to use the value of the energy gain of the water as the value for the energy of the cheese puff.)	Determine whether energy is conserved in the system of interest.
Compare the calculated energy content of the cheese puff with the food label energy content.	Account for the loss of energy. (The caloric value of the cheese puff is greater than the thermal energy increase of the water.)
Calculate the percent error. Account for this percent error.	—

The changes described for the revised versus traditional lab may seem trivial, and some will argue that the traditional lab also emphasizes the conservation of energy. To test that assumption, my colleagues and I "googled" the phrase "peanut lab calorimetry" (using this phrase because most teachers have used peanuts in the past) and then reviewed the first 23 sets of lab instructions (i.e., lesson descriptions and handouts) for this lab. Only seven of these sets of instructions mention energy conservation (one lab), conservation of energy (three labs), law of conservation of energy (one lab), or energy loss = energy gain (two labs). We found no other explicit phrases for conservation of energy (e.g., total energy remains constant). In the remaining 16 sets of instructions, the conservation of energy is implied and used but never stated. For example, one lab says, "The chemical energy stored in the peanut was released and converted into heat energy. The heat energy raised the temperature of the water in the small can." Notice that conversion is mentioned but not that one expects the same value for the energy before and after the conversion. Another lab says, "The peanut burned completely. It changed in mass. The water temperature was raised. When the peanut was burned, stored chemical energy was converted into heat energy, thereby raising the temperature of the water." Once again, conversion is emphasized, but the conservation of energy during this conversion is omitted.

Both of the examples quoted above from the 16 sets note that energy is transformed (i.e., chemical energy to thermal energy) and transferred (i.e., from peanut to the water), but both omit the crucial assumption that energy is conserved. The conservation of energy is implied but never stated in these examples. There is no explicit statement of the assumption that the energy gain of the water is *equal* to the energy loss of the peanut. Yet, the equivalence of energy gain and energy loss is crucial to the investigation and to science in general. The only reason we have the concept of energy in science is because it is conserved. It is the fact that energy is conserved that enables us to track its transfer, transformation, and dissipation within and between systems. If teachers are not mentioning energy conservation in their written materials, it may be assumed that in their instruction they are implying and making use of the conservation of energy but not explicitly focusing on this. By placing energy conservation in the foreground in the revised cheese puff lab, we emphasize the CCC and ensure that this important Big Idea takes center stage.

Lesson Plans for the Revised Cheese Puff Lab

In the revised cheese puff lab, the students are able to integrate the three dimensions of the *Framework*. Within science and engineering practices, they plan and carry out investigations. Within CCCs, they track the total energy in a system and emphasize that the total amount of energy in closed systems is conserved. Within DCIs, they show that "energy cannot be created or destroyed, but it can be transported from one place to another and transferred between systems."

Using the 7E instructional model (Eisenkraft 2003), my colleagues and I developed two example lessons: One can be adapted for use in upper elementary grades and the other for use in grades 6–12. We then used teaching expertise in different grade spans and the progressions in the *Framework* to differentiate the lessons to match grade-level expectations.

The lessons were mapped onto two tables (Tables 3.2 [p. 48] and 3.3 [p. 52]). The rows of each table are organized around six of the 7Es (Engage, Elicit, Explore, Explain, Elaborate, and Extend). The seventh E (Evaluate) takes place during all components of the lesson. The columns of each table summarize the components of the lesson, the connections to the *Framework* (i.e., science and engineering practices, CCCs, DCIs), and the evaluation and adaptations for different age groups (i.e., middle school, high school—physics, chemistry, biology).

After discussing all components of the lesson, we aligned the different components to the *Framework*. For example, during the Explore component, we discussed whether opportunities were provided for students to plan and carry out investigations or construct explanations and whether this was done in the spirit of the *Framework* document. Similarly, the lessons emphasize CCCs; obviously, energy as a CCC is crucial to this lesson, but Systems and System Models is another CCC that can be emphasized when the class discusses the energy losses to the environment during the experiment (see the "Explain" section of Table 3.3) and how transfers to the environment can account for the inability to show that energy was conserved in the calorimetry apparatus system (i.e., why the energy change of the water was less than the calorie content of the cheese puff). Finally, we identified connections to the DCIs at each grade level.

The *Framework* and *NGSS* place a new emphasis on Engineering Design as a DCI. Most students are not exposed to engineering principles and to the possible job opportunities for people with engineering degrees. This lesson can have Engineering Design concepts come into play during the Explain and Elaborate components in which students have to design a better calorimeter to minimize energy loss to the environment and compare their solutions with the actual apparatus used in food labs to derive the calorie values on packaging. Elementary school students can design containers with little energy transfer. High school students can explain why engineers use large volumes of water to conduct their calorimetry experiments. Specifically, why would a food engineer use a large volume of water and a relatively small temperature change as opposed to a small volume of water and a large temperature change? Energy efficiency and minimizing the transfer of energy and dissipation of energy is a constraint that many engineers deal with in design challenges.

No single set of lessons can adequately focus on all of the science and engineering practices, CCCs, or DCIs. In the revised cheese puff lesson plan, we chose to focus on a select few of each of these dimensions. For example, we emphasize energy conservation (in high school) and energy transfer but also discuss systems. In another set of lessons on a different topic, we may choose to emphasize the CCC of Patterns and mention the CCC of Cause and Effect: Mechanism and Explanation. In developing a curriculum for the year or for multiple years, all of the CCCs will be present at different times with varied emphasis.

Finally, in developing the two example lesson plans shown in Tables 3.2 (pp. 48–51) and 3.3 (pp. 52–54), we revisited the cheese puff lab from the perspective of how it must be adapted to the needs of different grade levels; specifically, which components of the 7E lesson should be modified and what additions and deletions should be made when the lesson is introduced in elementary school, in middle school, or in high school science classes. For example, the Engage component of the lesson will probably be different in fourth grade, middle school, and high school, where the students will have different levels of sophistication regarding the wellness concept, nutrition habits, and dietary needs. Similarly, in the Explore component of the lesson, fourth graders will be less advanced in their quantitative analysis. In contrast, high school students may go into much more detail of the bomb calorimeter that is used industrially in the Elaborate component of the lesson. In the Extend component of the lesson, middle school science students may focus on their life science studies as they discuss human digestion. The modifications are guided by the *Framework* presently but can be further guided by the *NGSS*.

Table 3.2.

LESSON PLAN FOR ELEMENTARY SCHOOL USING THE 7E INSTRUCTIONAL MODEL

Components of the lesson	Framework connections (a few examples)	Evaluation (checking for understanding)
ENGAGE: *In the book* Bread and Jam for Frances,* *Frances eats bread and jam for breakfast, lunch, and dinner. She decides that, as much as she likes bread and jam, other foods are good to eat also. Is there one food you would want to eat every day for every meal?*		Do students name their favorite foods? Do they differentiate between healthy and junk foods?
ELICIT: *When we cook, we often need to heat up water. What are some of the ways you have heated up water or seen people heat up water?* This question will help elicit students' understanding. Make the connection between ENGAGE (favorite foods) and methods of heating water more explicit. For example, if no student mentions a favorite food that is boiled in water, the teacher can share his or her own favorite food that requires heating water, such as macaroni and cheese or ravioli. Then, begin the ELICIT by discussing ways to heat that water. Students may respond by naming technologies like stoves and microwaves. Listen intently to find out if any students think about the flame or fire as the source of heat. *Another way we can heat up water is to make a hot fire and place the water over it.* If a student has already mentioned fire or flame, skip this sentence. *What do we need to make a fire?* Listen for students' responses. The goal is to help them articulate that combustible raw materials, like logs, are needed for a fire to burn. *Do you think that we can use cheese puffs as the material to burn for a fire?* They may think that is a funny idea and be curious about it! *We're going to try to burn a cheese puff today. And we will see if we can use that cheese puff fire to heat up water.*	Practices: Asking questions and defining problems Disciplinary core ideas: Energy in chemical processes and everyday life	Do students mention appliances? Do any students mention the flame? Does anyone mention heat?

Table 3.2 (*continued*)

Components of the lesson	Framework connections (a few examples)	Evaluation (checking for understanding)
EXPLORE: Part 1: *Today we are going to investigate how good a burning cheese puff is at heating up water. We need to set up supplies so that the cheese puff can burn and, while it is burning, it heats up the water. Can you think of a way we could use these materials* (present the materials needed for the investigation) *to set up our investigation of how well a burning cheese puff heats up water?* Students will have to generate ideas for holding the cheese puff and the can or glassware with water. You can also show them alternatives for holding the cheese puff and ask which way they prefer. Typically, the cheese puff is placed on an unfolded paper clip or held with a straight pin with one end pushed into a cork. The beaker of water is placed above the cheese puff. *How will we know that the water has gotten hotter?* Students should suggest using a thermometer. After deciding on a setup, the experiment can be performed. The students can see the cheese puff burn and measure the change in temperature of the water using a thermometer. Since burning the cheese puff is a safety hazard for students, the teacher can burn the cheese puff and the water can have a few thermometers in it for student groups to read the values. Alternatively, for a demonstration, a thermometer with a large digital display might work best. The data can be presented in a graphical display (e.g., chart or bar graph). Part 2: *What could be changed in the experiment so that the temperature of the water will increase even more?* Students should generate ideas such as increasing the number of cheese puffs and decreasing the amount of water. An alternative suggestion is to use a different type of food such as a potato chip or vegetable. At this point, students will have made a claim: To increase the temperature of the water, you should burn more cheese puffs. You may choose to help students test this claim by repeating the experiment with three or five cheese puffs. Similarly, you may choose to help students test the other claims regarding the amount of water or the use of a different type of food.	Practices: Asking questions and defining problems; planning and carrying out investigations; engaging in argument from evidence	Students can describe or draw a setup. You can ask other students to come up with alternatives. All students should be able to think of a stand for the cheese puff. Do students read the thermometer correctly? Are they able to make the connection between the thermometer reading and how hot the water is? Students should explain why they think five burning cheese puffs will raise the temperature more than one burning cheese puff.

Table 3.2 (*continued*)

Components of the lesson	*Framework connections (a few examples)*	Evaluation (checking for understanding)
EXPLAIN: Students should use their experimental evidence (e.g., a comparison of the temperature change of the water with one burning cheese puff vs. five burning cheese puffs) to support their original claim. You can introduce the notion that energy released as the cheese puff burned went to heating the water. Students should then rephrase their claim and evidence using words such as *energy* and *heat*. Connect back to the original ELICIT question: *Remember that we asked how can we heat water? Would you agree that we have found a way to use cheese puffs to heat water? What have we discovered about how to heat water with cheese puffs?* Students can reiterate their claims and evidence here. *Now, do you think other foods besides cheese puffs could be burned to heat up water?*	Practices: Analyzing and interpreting data; engaging in argument from evidence; obtaining, evaluating, and communicating information Crosscutting concepts: Energy and matter; cause and effect Disciplinary core ideas: Definitions of energy	Students should be able to explain their results using a claims-evidence format. Students should predict that if 10 cheese puffs were used, the water temperature increase would be even greater.[†]
ELABORATE: *Our investigation tells us something about what food does for us. When we burned the cheese puff, it heated up the water. When we burned more cheese puffs, they heated up the water even more!* *We eat food, like cheese puffs. We eat it because it tastes good and because it keeps us healthy (though some foods are healthier for us than other foods.) One way the food we eat keeps us healthy is by keeping our bodies warm.* Students should try to imagine how the cheese puff (or any other food) heats up their body. *Is it like the cheese puff burning? Can the cheese puff burn so slowly that there is no flame? How could we find out?*	Practices: Constructing explanations	Students should recognize that "flames inside the body" would be hazardous.

Table 3.2 (*continued*)

Components of the lesson	*Framework connections (a few examples)*	Evaluation (checking for understanding)
EXTEND: *Since burning cheese puffs produces a different amount of heat than burning cashews, how would you rank foods so that we know in advance which foods we want to eat and which foods we don't want to eat?*	Practices: Asking questions	Some students may know about food labels and calories. Where have they heard about calories? Students may be able to discuss their personal theories of what makes some foods healthy foods and some junk foods.
EVALUATE: Please see evaluations for each component of the 7E model in the right-hand column.		

Note: In the "Components of the lesson" column, text in italics is meant to be spoken by the teacher to the class.

* *Bread and Jam for Frances* is a book by Russell Hoban (pictures by Lillian Hoban) published by HarperCollins, 2008; the book is for grade levels kindergarten through grade 3.

† A caveat is that if the sample of water is small enough, using more cheese puffs can easily start the water boiling, which makes things more complicated. When something boils, its temperature actually stays the same as more thermal energy is transferred to the water—for reasons too complex to go into with students at this age.

So, if you try adding cheese puffs, be sure that you are heating enough water that it doesn't begin to boil.

Table 3.3.

LESSON PLAN FOR MIDDLE SCHOOL AND HIGH SCHOOL USING THE 7E
INSTRUCTIONAL MODEL

Components of the lesson	Framework connections (a few examples)	Adaptations for different grades (a few examples)
ENGAGE: Read the calorie labels of a few different food products to the students. You can also ask students if they have ever gone on a treadmill, mentioning that some treadmills inform the person exercising that every 20 minutes of exercise burns 100 calories. *When people decide to eat an extra snack, do they compare the calories of the snack with the exercise required to make up for the additional calories?*	Disciplinary core idea: Definitions of energy	Relate this to the wellness concept.
ELICIT: *How do the food suppliers know the calorie count of the foods that they display on the label?* This question will elicit students' prior understanding of the concepts in the lesson.	Practices: Asking questions and defining problems	Rank three food products in order of calories per serving.
EXPLORE: *In this lesson, we will use the calorie content of the foods as displayed on the package to determine if the energy loss of the food (cheese puff) is equal to the energy gain of water being heated.* This may require a mini-lesson to assist students in understanding calories per gram and calories per serving. The first "Explore" of this lesson can be a qualitative exercise where students compare the temperature change of water heated with one cheese puff or three cheese puffs. The second "Explore" can be a quantitative exercise where students measure the temperature change of the water. The research question, *Is energy conserved in our apparatus system?* will require students to calculate the energy gain of the water from its temperature change and compare this with the energy loss of the cheese puff, using the measured change in mass of the burned cheese puff and the energy in calories of that amount of cheese puff from the food label.	Practices: Asking questions and defining problems; planning and carrying out investigations; using mathematics and computational thinking	How do we connect temperature, heat and calories? Limit the experiment to the idea that increased mass of cheese puff burned is related to an increase of temperature (qualitative exercise only). High school: What are the uncertainties of each measurement?

Table 3.3 (*continued*)

Components of the lesson	Framework connections (a few examples)	Adaptations for different grades (a few examples)
EXPLAIN: Students discuss how to calculate E_c (energy transferred from the burning cheese puff) and E_w (energy transfer to the water). Conservation of energy in our apparatus system requires that $E_c = E_w$. In the experiment, all groups will probably find that $E_c > E_w$. *Energy is not conserved in this system!* *How can we explain why energy was not conserved in the experiment we conducted?* Students must now consider possible explanations for energy not being conserved: *Explanation 1: The conservation of energy as a principle of science is incorrect.* As part of the classroom discussion, you will have to explain that this is a bedrock principle of all science. *An organizing principle of energy conservation cannot be discarded on the basis of a single 30-minute experiment. We will have to assume that it is true and come up with another, more plausible explanation to explain the results of our experiment.* *Explanation 2: The food label was incorrect and we therefore used an incorrect figure for E_c, the energy content of the cheese puff.* Some food labels have been found to be incorrect and have led to lawsuits, a societal topic that students can further investigate. Since all cheese puff food packages give comparable numbers, it is safe to assume that these packages probably are not all incorrect. *Explanation 3: We made an error in the calculations.* Students can check their calculations as well as checking the validity of the equation. Does the equation $Q = mc\Delta T$ make sense where Q is the thermal energy, m is the mass of the water, c is the specific heat of water and ΔT is the change in temperature? Is more thermal energy required for more water for the same temperature change? Is more thermal energy required for a bigger change in temperature of the water? Explanation 3 as a claim is not supported by the evidence. *Explanation 4: The energy transferred from the burning cheese puff is less than the energy transferred to the water because of the poor design of the apparatus, resulting in large heat loss to the air, to the beaker, and to the surroundings.* This discussion also emphasizes the crosscutting concept of systems. The energy loss occurs because the cheese puff and water are not an isolated system.	Practices: Using mathematics and computational thinking; engaging in argument from evidence; obtaining, evaluating, and communicating information Practices: Constructing explanations and designing solutions Crosscutting concepts: Systems and system models	How do you know that the cheese puff has energy? How is energy transformed? How would the temperature of the water change (a) if we used less water in the experiment and (b) if we used more cheese puffs? What forms of energy are present? High School: How does belief in the conservation of energy relate to the nature of science?

Table 3.3 (*continued*)

Components of the lesson	Framework connections (a few examples)	Adaptations for different grades (a few examples)
ELABORATE: Students suggest improvements in the design of the laboratory investigation. They can explain how each improvement will minimize the energy loss. As a further "Elaborate," students can compare their experimental design with that of food scientists who use a bomb calorimeter. This research regarding food science can lead to a discussion of why the scientists use a very large amount of water, which produces a very small temperature change in the water.	Practices: Planning and carrying out investigations Disciplinary core idea: Engineering design	Relate changes in the experiment to the engineering design cycle.
EXTEND: The cheese puff lesson can now be extended to discussions of molecular bonding, combustion reactions, energy content of different foods, human digestion, specific heat, and other topics related to the curriculum that the students are studying.	Disciplinary core ideas: Energy in chemical processes and everyday life; organization for matter and energy flow in organisms	Burning (digesting) food provides energy. How does this lab relate to our study of digestion? High School: Can the energy content of the cheese puff be determined from an analysis of the combustion reaction?
EVALUATE: The evaluation takes place during all components of the lesson (Engage, Elicit, Explore, Elaborate, Extend). A test can also be constructed in which students have to measure the energy content of another food product and answer questions on the material in the lesson. Some of these questions should emphasize the crosscutting concept of energy conservation.	Practices: Planning and carrying out investigations; constructing explanations and designing solutions	

Note: In the "Components of the lesson" column, text in italics is meant to be spoken by the teacher to the class (with the exception of italicized variables).

Tips for Selecting and Adapting Lessons

Part III of the *Framework* is entitled "Realizing the Vision," and the first chapter in Part III is entitled "Integrating the Three Dimensions." This chapter of *Teaching Energy Across the Sciences, K–12,* has shown how the three dimensions can be integrated in the cheese puff lab investigation, and this example can serve as a model for integrating the three dimensions in other lessons (see Table 3.4).

Table 3.4.

SELECTING AND ADAPTING LESSONS TO INTEGRATE THE THREE DIMENSIONS OF LEARNING

Step	Application of the general steps to the traditional cheese puff lab as described in the chapter
Choose a lesson that is already part of the school curriculum.	We chose the cheese puff lab because it can emphasize different concepts at different grade levels. In the early grades, the lesson may focus on students learning that many cheese puffs have more energy than a few cheese puffs and that we eat cheese puffs to keep our bodies warm and our muscles working. The lesson may also focus on how different foods have different calories. The lesson may also emphasize the wellness concept. In the higher grades, calculations of energy are appropriate, as well as more sophisticated questions regarding engineering design.
The lesson can probably include many of the components of the *Framework.* Use the *Framework* as a guide to see which components can fit with the lesson.	The students can be involved in science and engineering practices by first planning and carrying out an investigation in which they help design the experiment in which a cheese puff is burned and water is heated. They can ask questions (for science) such as how many cheese puffs must burn to heat the water to a certain temperature. They can define problems (for engineering) by investigating the best way to hold the cheese puff while it is burning or how to design an apparatus that minimizes heat transfer to the surroundings.
Identify the crosscutting concepts that frame the lesson and choose one as the primary focus.	The primary focus is on the conservation of energy—did all of the energy transferred from the burning cheese puff result in an equal gain in energy of the water as it heated up?
Restructure the lesson so that one of the crosscutting concepts is front and center rather than being implied, but not stated, in the lesson.	The research question needed to be altered to emphasize the conservation of energy rather than emphasizing the calculations. The crosscutting concept of energy conservation becomes the focus, as opposed to determining the calories in a cheese puff by assuming conservation of energy.

Table 3.4 (*continued*)

Step	Application of the general steps to the traditional cheese puff lab as described in the chapter
Use the 7E or some other instructional model to ensure that practices are integral to the lesson.	Using the 7E model reminds teachers planning the lesson that the Engage component comes first and that the Explore component occurs before the Explain component.
In developing the lesson, refer back to the Framework to identify where each of the three dimensions, including the disciplinary core ideas, appear in the lesson.	Science and engineering practices, crosscutting concepts, and disciplinary core ideas are all woven together and are explicitly denoted in the plan.
After you are satisfied with the quality of the lesson, participate in a vertical teaming exercise to see how modifications of the lesson will take place in different K–12 grades.	In the revised cheese puff lab for elementary school, students can ask how the temperature increase of the water may change with different amounts of cheese puffs. Middle school students can measure the temperature change of the water (as do the elementary students) but can also complete the calculations of the corresponding energy gains of the water. High school students may use a calorimeter and track the losses of energy from the cheese puff–water system. By intentionally focusing the investigation in ways that correspond with the practices, crosscutting concepts, and disciplinary core ideas emphasized in each grade level, students are well positioned to develop more sophisticated understandings of complex phenomena (such as the cheese puff lab) and nuanced concepts (such as energy) over time.

Summary

Energy may be the most important concept in all of science. As discussed in Chapter 2, defining energy may be a futile exercise. But by focusing students' attention on what energy *does* rather than what it *is*, we can help them develop an understanding of how to use energy to make sense of a wide range of complex and interrelated systems.

In the revised cheese puff lab, we can draw students' attention to different forms or manifestations of energy (Big Idea 1), energy transformation (Big Idea 2), energy transfer (Big Idea 3), energy conservation (Big Idea 4), and energy dissipation (Big Idea 5). Our research question pertains to energy conservation and provides an opportunity to better understand this concept. Energy is not conserved in the apparatus system (as evidenced by the experiment), and students must confront the concept of energy transfer and dissipation to the surroundings. The students can recognize that energy transfer occurs from the cheese puff–air system to the water and container system. Energy transformation is apparent as the chemical energy associated with the cheese puff and oxygen in the air manifests itself as thermal energy of the water.

Although the revised cheese puff lab focuses on energy as a CCC, we can choose to connect this lab to a range of more discipline-specific investigations. For example, we can use the Five Big Ideas about energy to relate the burning of the cheese puff to the process of human digestion when we eat food or to energy transformations and transfers occurring in connection with nuclear fusion in the Sun.

By consistently and explicitly emphasizing the Five Big Ideas of energy, science instruction can put students in a much better position to reliably use the energy concept across a wide variety of disciplinary contexts. This is the power of crosscutting concepts in science.

References

American Association for the Advancement of Science (AAAS). 1993. *Benchmarks for science literacy*. New York: Oxford University Press.

Eisenkraft, A. 2003. Enhancing the 5E model. *The Science Teacher* 70 (6): 56–59.

National Research Council (NRC). 1996. *National science education standards*. Washington, DC: National Academies Press.

National Research Council (NRC). 2012. *A framework for K–12 science education: Practices, crosscutting concepts, and core ideas*. Washington, DC: National Academies Press.

NGSS Lead States. 2013. *Next Generation Science Standards: For states, by states*. Washington, DC: National Academies Press. *www.nextgenscience.org/next-generation-science-standards*.

SECTION 2

TEACHING ENERGY ACROSS THE LIFE, PHYSICAL, AND EARTH SCIENCES

CHAPTER 4
TALKING ABOUT ENERGY

JEFFREY NORDINE

In science teaching, we are used to thinking about language in the context of helping students learn complex scientific terms that they do not use in their everyday lives. You or your colleagues may post science-specific terms like *covalent, scree,* or *hadron* around the classroom to help students learn such unfamiliar words. It can be a challenge to help students remember the meaning of esoteric science terms, but when teaching about energy we often have a different challenge—helping students recognize when to connect a specialized definition to widely used words. In energy instruction, everyday words like *conservation, work,* and *potential* take on special meanings that often do not align with how they are used in everyday conversation. Additionally, some energy terms commonly used by scientists, such as *flow* or *forms,* may imply to learners that energy has characteristics that it does not have. This chapter is dedicated to unpacking the language that scientists use to discuss energy and identifying guidelines for using energy-related words as students develop their understanding of energy over time.

Why Language Matters

Language affects how students participate in classroom science (Lemke 1990), but even more fundamentally, language mediates the very ways that we think. Lev Vygotsky (1978) wrote about the importance of language as a cultural tool that affects how humans engage in the construction of knowledge. This goes both for how we talk to one another and how we talk to ourselves. By mediating how we talk and think, language affects our understanding.

If you're a native English speaker, chances are that your inner narrative happens in English. Using English words to frame your thoughts means that if there is an English word for a concept, it is easier to think about—if there is not, it can be difficult to recognize that idea. If you have ever learned another language, you may have noticed times that a word in the new language captures a sentiment or an idea that is difficult to express in English. For example, Germans use the word *Treppenwitz* to describe a joke or comeback that you thought of just a little too late. In English, the word literally translates to "staircase wit," but the German word describes that experience of thinking of a great comeback or remark after the moment for it has passed—say, on the staircase on the way out the door.[1] Native English speakers are very familiar with this idea, but having a word for this experience unifies the feeling and makes it easier to think and talk about it.

Another situation that illustrates how language affects understanding is when we use the same word for multiple ideas; it may be difficult to distinguish between the meanings unless they are in context. Take, for example, the word *trim.* If you tell someone you are going to trim the tree, this can mean very different things depending on whether you are holding a pair of hedge clippers or a box of holiday decorations.

Scientists name ideas to make it easier to talk about them. Sometimes, this means inventing words like *chromatography* or *asthenosphere.* These words typically have little meaning in a nonscientific setting, but they are critical for helping scientists think and talk efficiently about the ideas the words represent. Sometimes, however, scientists borrow words that are already in common use and give them a second, scientific meaning that accompanies their everyday meaning. Robert Hooke borrowed the word *cell* to describe what he saw when he looked at plants under the microscope. This word was common before Hooke used it to describe the basic functional unit of living organisms, and it has retained several original meanings. In addition to its scientific meaning, *cell* can mean a small room for inmates, a small group of people, or a mobile phone. In general, however, these meanings are highly context dependent and don't interfere with one another; very few students have ever thought that organisms are made up of tiny phones.

Like *cell, energy* is also a borrowed term, but unlike *cell, energy* has meanings in different contexts that have substantial overlap. These overlapping meanings can contribute to students' confusion about energy as a scientific idea, and the reason why there is such overlap in meanings has to do with how its meanings have evolved over time.

A Brief History of the Word *Energy*

Do an internet search for the word *energy* and you will see plenty of articles that discuss energy as a scientific idea, pictures of technological devices, and news stories about the cost of a barrel of oil. In today's world, energy is a scientific and technological idea. But it hasn't always been like that.

It surprises many people that the word *energy* was actually used in everyday language long before it became a scientific term. Thomas Young, an English physicist, decided to borrow this word, which was already in common usage to describe strength and vitality, to refer to what we might now call mechanical energy (Coopersmith 2010). The word can be traced back to the ancient Greek term ευέργεια (transliterated as "energeia"), when it

---

1 German speakers may also know that, over time, this word has taken on another meaning that is closer to "irony" or "a silly joke or behavior."

was used to describe activity or vigor. The meaning of energy as a general sense of activity, strength, and vigor goes back thousands of years, but it wasn't until well into the 19th century that *energy* was used as a scientific term. Scientists appropriated this word and now unfairly criticize the public for using it in ways that are different from the precise scientific meaning. Nowadays, a student in science class may be corrected if he says something such as "drinking caffeine gives me energy"; his teacher may be quick to point out that caffeine contains no calories and could not possibly have given him energy. But the student *feels* more vital, more energetic after consuming caffeine, so is he wrong for using the everyday, historical meaning of the term?

In some ways, what happened to the term *energy* is similar to what has happened more recently to the word *spam* in the era of e-mail and social networking. Spam was originally a trademarked name for a canned meat product, but it has acquired a very specific new meaning in the computing era (i.e., to indiscriminately distribute an unsolicited message). Likewise, before social networking, a *tweet* used to mean only the sound that a bird makes and *friend* used to be exclusively a noun. Just as one could hardly fault your grandmother for thinking Spam is food or that a tweet is an avian noise, it is equally understandable that prior to instruction, a person might conceptualize the word *energy* as a type of human vigor or vitality, because that is what the word originally meant.

The problem with the historical meaning of the word *energy* is that it is actually quite close to its scientific meaning. Our feeling of human strength and vitality does depend on the scientific concept of energy—to run fast, jump high, or push hard, our bodies need to be able to convert chemical potential energy into other forms. But energy as a scientific idea is not exclusively a property of humans or living organisms, and students often enter science classrooms with the belief that energy only applies to living things. Holding the idea that energy is only related to living organisms can make it difficult to develop a more complete understanding of energy.

The statement "drinking caffeine gives me energy" is not a problem if we are only concerned with the historical meaning of the word *energy*. But if we are trying to teach students about energy as a scientific idea, this statement may convey the idea that caffeine carries energy, which it does not (caffeine cues a set of metabolic effects that facilitate the body's ability to convert chemical potential energy that it has already stored). If the student says instead that "drinking caffeine corresponds with my feeling energetic," then his teacher may not be so quick to correct him. Word choice and phrasing matter in science teaching, but they especially matter in energy teaching because there are so many common words associated with the energy concept that have meanings that are close to—yet distinct from—their everyday meanings.

Many different people talk about energy in lots of different ways. This is true across scientific and everyday contexts, and it is true across scientific disciplines. If we are to help students build an understanding of energy that connects to their intuitive ideas and is useful as a crosscutting concept, then we need to be careful about the language we use to talk about energy within disciplinary contexts. In the next sections, we discuss several key words in energy instruction that may promote student confusion and incorrect ideas about energy if we are not careful about how we use them.

Talking About the Five Big Ideas

Focusing on the Five Big Ideas in energy instruction (as described in Chapters 1 and 2) can help clarify the most important ideas about energy and how it is used across disciplines. These Five Big Ideas can simplify the energy concept for students by providing simple and consistent ways to talk about energy in a variety of contexts. However, the words we use to discuss these ideas carry connotations that may imply things about energy that are not true. Although there is no perfect solution to avoiding confusion, an important start is to clarify for ourselves what the terms related to the Big Five Ideas mean and to identify how they may potentially be misinterpreted. In this way, we can be more clear and consistent when talking about energy with students.

Energy Forms

Energy does not have a form. It is not a substance and does not have any material characteristics, thus, energy does not "take form." We cannot see it or touch it. In fact, we can never even measure energy directly! Every time we speculate on how much the energy of a system has changed, we do so by performing calculations based on observable characteristics of the system, such as speed, height, stretch, temperature, or mass.

We do not observe kinetic energy directly; instead, we observe speed relative to some reference frame. We calculate a value for kinetic energy using the mathematical formula

$$kinetic\ energy = 1/2 \times (mass) \times (velocity)^2$$

Likewise, we do not observe gravitational potential energy; instead, we observe the separation between objects and the Earth relative to some reference separation. For example, we can calculate a value for the potential energy between an airplane and the Earth using the mathematical formula

$$gravitational\ potential\ energy = (mass\ of\ the\ airplane) \times$$
$$(acceleration\ of\ Earth's\ gravity) \times (height\ of\ the\ airplane\ above\ the\ ground)$$

You may have seen or used formulas for calculating the energy of a stretched spring ($E = \frac{1}{2}kx^2$), the energy associated with a photon of light ($E = hf$), or the energy associated with mass ($E = mc^2$). Each of these formulas gives us a way of translating the measurements we make about systems into a numerical value for energy, and it is these numerical values that allow us to use the conservation of energy to set limits on the behavior of systems.

Over time, scientists have developed a shorthand for discussing each of these ways of calculating values for energy—forms. What we call forms of energy roughly corresponds with a different way of calculating numerical values for energy. When we use the formula $E = \frac{1}{2}kx^2$ to calculate energy, we call it elastic energy. When we use the formula $E = hf$, we call it light energy. The formula $E = mc^2$ helps us calculate what we call mass energy. When using the term *energy form,* a scientist is really letting you know what characteristics of a system are changing and what formulas are useful for quantifying energy changes associated with phenomena or processes.

Scientists may agree that energy forms are really just different ways of calculating energy, but young students have not yet learned this, and the term may imply that energy has a physical form. Some science educators have attempted to sidestep this issue by using other words, such as energy *type,* but this word may imply that there is more than one kind of energy, when, in fact, all energy is fundamentally the same thing. Whether we use *form, type, store,* or *manifestation,* there is no perfect word, but by far the most commonly used word—and what you will see in the *Next Generation Science Standards* (*NGSS*) performance expectations (NGSS Lead States 2013)—is *form.* Even though energy forms are not precisely defined (Quinn 2014), the term makes it easier to talk about the role of energy in phenomena using everyday language.

When using the term *energy form* in the classroom, we need to be aware that this term carries a connotation that energy is some material substance that can take on various physical characteristics. By doing activities and facilitating discussions that reinforce that all forms of energy can change into one another and, thus, are fundamentally the same thing, we help students understand that this term is just an expression for describing all of the ways that energy changes can be tracked by observing the measurable characteristics of a system.

Energy Transformation and Transfer

When watching a phenomenon or process unfold, we notice that any time one form of energy increases, at least one other form must simultaneously decrease. Because it seems that one form is becoming another, we call this process energy transformation. In the *NGSS,* you will see the terms *transformation* and *conversion* used to describe this process. Both of these terms, however, are distinct from the term *energy transfer.*

Transformation and conversion refer to energy changing from one form to another; energy transfer refers to energy crossing the boundary between systems or objects. For example, energy can be transferred from one marble to another when they collide. As a result of this collision, if the kinetic energy of one marble increases and the kinetic energy of the other decreases, then we refer to this process as an energy transfer because energy crossed the boundary between marbles. If you look closely, this transfer process involves a transformation, as the kinetic energy briefly became elastic potential energy when the marbles were in contact and were slightly deformed from their original shape. Transfer processes often involve energy transformations, but they don't have to. If we define our system as a region of space (e.g., the Earth and the region up to 100 km above it), then energy can enter or leave our system as matter enters or leaves. If an oxygen molecule in the upper atmosphere happens to move away from the Earth without striking another molecule, it will leave Earth and never return, taking its energy with it.

Although energy transfers nearly always involve transformations, these two terms mean different things. Energy transformations are processes that convert energy from one form to another, whereas energy transfers are processes that carry energy across system boundaries.

Just as we can identify a variety of energy forms (e.g., kinetic, thermal, elastic), we can likewise categorize a set of energy transfer processes. When energy is transferred via force, we call this work. Energy can also be transferred via sound or light waves (in which energy transfers between systems but matter does not), electricity (in which an electric field transfers energy by moving charged particles), or heat (in which energy transfers from a hot object to a colder one).

Energy Dissipation and Loss

As heat transfers energy from regions of high temperature to low temperature, particles that were moving fast tend to slow down and particles that were moving fast tend to speed up, until all particles are moving with about the same kinetic energy. In this way, thermal energy is evenly distributed throughout a system in a process called dissipation.

Dissipation can be a confusing word for students because it implies that energy vanishes—that it is, in fact, gone. If a small hot system interacts with a larger cool system, thermal energy will transfer via heat from the small system to the larger system until both come to the same temperature. If the large system is large enough, the thermal energy transferred to it from the smaller system will have virtually no effect on its temperature. If you poured a pot of boiling water into the ocean, there is no way you could measure an increase in the ocean's average temperature around the globe. There is a technically a change in the ocean temperature, but it is so slight that you could never measure it. Is the thermal energy from the boiling water gone? No! But it has dissipated over such a vast region of space that you could never hope to measure it or recollect and reconcentrate it. The thermal energy is effectively lost.

When scientists say that energy is lost, they mean that it has been transferred as thermal energy to such a large system that one cannot hope to recover it again. In this way, energy loss is actually very close to how we use the word *loss* in an everyday setting. If a ship is lost at sea, we accept that it is out there somewhere, but in an expanse so vast that hope of ever recovering it is all but gone. Similarly, energy still exists when it is lost, but hope of recovering it for useful purposes is gone.

Energy Conservation and Use

In everyday language, *conservation* is an environmental or ethical choice. Children encounter this term in a variety of settings, in the context of water, wildlife, energy, and even art conservation. As an environmental or cultural choice, the term *conservation* refers to keeping something in its original condition.

In science, the term *conservation of energy* is a fundamental statement about the nature of energy: that it is never created or destroyed. Yet, even though energy is never destroyed, it can be more or less accessible for doing things such as running machines. Its accessibility is based on a number of factors, such as how spread out it is within a system.

When we say that we use energy, this means that we transfer energy into a system or device (e.g., our car) in a form that is concentrated and readily available. In the device, the energy is transformed as the device operates. For example, we transfer energy to our cars via gasoline, which readily reacts with oxygen when it is heated and releases a lot of energy per kilogram burned, some of which can be harnessed and transformed into kinetic energy of the car. Gasoline is a very convenient way to power cars because it is so energy dense and its energy is so easily accessible. When we use the energy in gasoline to power our cars, its energy is not gone, but it is lost to the Earth's environment and outer space as thermal energy. The natural resources (such as crude oil) used to produce the gasoline, however, are gone. The oil molecules that stored so much chemical energy are broken apart and rearranged into lower potential energy configurations, and these molecules will not spontaneously reform. When we say that we should drive less to conserve energy, what we really mean is that we should conserve the energy resources that went into producing gasoline—that we should maintain these energy resources in their original condition.

It is worthwhile to discuss both *conservation of energy* and conservation of energy *resources* in the science classroom, but we should clearly and consistently use the correct term for each.

Other Potentially Confusing Energy Terms

Outside of words related to the Five Big Ideas about energy, there are a variety of terms common in energy instruction that can promote incorrect ideas about energy and how it behaves.

Work

As we mentioned in Chapter 2, it is common to try to define energy as the ability to do work, even if one has not defined what work is. Students commonly accept this definition even without a formal scientific definition of *work*, because we all have an intuition of work as something that requires effort. But work is a precisely defined scientific quantity that describes how much energy is transferred between systems via force. It is possible to push on something and do no work on it; for example, a bookshelf exerts an upward force on a book, which counteracts the force of gravity, but this force doesn't transfer energy to the book and thus it does no work.

When we speak about work in science classes, it is important to remember that we are describing a *transfer process*—that is, we must specify an origin and a destination of this energy transfer. If we are describing the transfer of money between bank accounts, it is very important to specify the account from which the money should be transferred and the account to which the money should be transferred. If your bank sent you a confirmation that your account had a $1,000 transfer, you'd probably want more information! To or from my account? From where? To where? The same is true for work. We need to do much better in specifying which systems are involved, which one received the energy, and which one transferred it away.

In many physical science classes, we love to have a student volunteer stand in front of class holding a heavy book at rest and proudly declare that the student is doing no work since she is not moving the book. After all, work can be calculated by the formula *work = force × distance*,[2] and if she isn't moving the book, then work equals zero. This declaration surprises and amazes students, and it is also wrong—or at least imprecise. Considering work as a transfer process, the statement "she is doing no work" fails to identify an origin and destination system. A more complete statement is that "she is doing no work on the book," because it conveys the systems of interest in this statement. Even better, the statement "the force exerted by the student on the book is doing no work on the book" identifies the origin system, the target system, and the force in question.[3]

It is tempting to ignore that work is fundamentally a process by which a force acts to transfer energy between systems and, instead, simply have students calculate values for work, since the equation is so straightforward. But focusing on the equation, rather than having students use words to formulate a complete description of the energy transfers that occur when forces do work, ignores the core of what work is and why we care about it.

By talking consistently about work as a transfer process and specifying an origin system, a destination system, and the force of interest, students are in a better position to understand connections between energy transfer and force. Further, they are more likely to see that defining energy as the ability to do work is essentially saying that energy is the ability to transfer energy.

---

2 Of course, this formula is a simplification of the more general formula for calculating work and is only applicable in certain situations. The full equation is more general, but this is the form most commonly presented at the middle school level.

3 Although this is a common demonstration that illustrates how the scientific meaning of work is different from our everyday use of the term, it is not technically accurate to say that the student is doing no work on the book in such a demonstration. There are many micromovements as the student is holding the book steady, and there is a transfer of energy to the book in these movements. Furthermore, for the student to keep her muscles contracted, chemical reactions must take place within her body that transform and transfer energy, making the student feel tired.

Potential Energy

Many students think that the term *potential energy* means that something has the potential to have energy. That is, students may think that potential energy is not really energy but rather some kind of dormant form of the real thing. In some sense, *potential energy* is an unfortunate term, and it caught on before scientists agreed that potential energy is every bit as real as kinetic energy (Hecht 2003). It's probably safe to say that if scientists had known then what we know now, potential energy would have a different name, like position energy or configuration energy, but it looks like we're stuck with the term *potential energy* for the foreseeable future.

The reason why position or configuration energy would be better names is that potential energy is associated with the arrangement of mutually interacting bodies. Gravitational potential energy is associated with how far apart an object is from Earth. The farther the object is from Earth, the more potential energy exists in the Earth-object configuration. The potential energy exists in this situation because all massive objects exert a force of gravity on each other. Earth pulls downward on a skydiver, and the skydiver pulls up on the Earth with an equal and opposite force of gravitational attraction. As a skydiver begins to fall, the potential energy of the skydiver-Earth system begins to be converted into kinetic energy of the objects in the system. Yes, *objects*! Earth moves, too, but while the force between Earth and the skydiver causes a noticeable change in speed for the 50 kg skydiver, this force is far too small to cause a measurable change in speed for the 5,970,000,000,000,000,000,000,000 kg Earth. So, although it seems like the only object in this system gaining kinetic energy is the skydiver, rest assured that Earth does, too. Just like we could never measure a change in the average global ocean temperature if someone poured a pot of boiling water into it, we will never notice a change in Earth's motion if someone jumps out of an airplane.

Like gravitationally attracting objects, magnets store potential energy due to their configuration. If you bring a north pole and a south pole of two small magnets close to each other, you will almost certainly notice that the magnets pull on each other. If you let them go, you will notice that both magnets start moving toward each other. They are converting potential energy associated with their configuration into kinetic energy as they move. If you instead use one really large magnet and a small one, you will notice that the small one moves more than the large one as they come together. As one magnet gets more and more massive compared with the other, this situation begins to more closely resemble the case of a skydiver falling toward Earth.

For both the magnets and the skydiver, the potential energy of the attracting objects increases as the objects are farther apart—the higher you drop an object above Earth, the faster they smack together. For mutually attracting objects, more distance between them corresponds to more potential energy. The opposite is true for mutually repelling objects, such as two magnetic north poles or two negatively charged objects. If things repel each other, then the potential energy associated with their arrangement increases as they get closer to each other. The closer they are in their initial arrangement, the more rapidly they will fly off from each other when released.

Whether related to attractive pulls or repulsive pushes, potential energy is fundamentally connected to force. There are many varieties of potential energy, and every type of potential energy is connected to a force. Gravitational energy, or gravitational potential energy, arises because of the force of gravity between objects. Chemical energy, or chemical potential energy, arises because of the electric force between charged atomic nuclei and electrons. Nuclear energy, or nuclear potential energy, arises because of the (very strong) forces holding an atomic nucleus together. Anytime the value of one of these potential energies increases, the interacting objects must have less kinetic energy (move slower); when the potential energy of a system decreases, its constituents move faster.

Potential energy is the counterbalance to kinetic energy, and it is every bit as real as kinetic energy. Further, potential energy is fundamentally associated with systems of objects and never a single object by itself. Our language in the classroom should reflect that potential energy is real and associated with systems. Phrases such as "a boulder on a cliff has a lot of potential energy" fail to convey that the potential energy is really stored in the boulder-Earth system; by focusing only on the boulder in this situation, our language fails to focus students on the key aspect of potential energy—its relationship to the configuration of a mutually interacting *system* of objects. Saying something such as "a boulder on a cliff is a high potential energy configuration of the boulder-Earth system" is a much better characterization of what we—somewhat unfortunately—call potential energy.

Energy in Chemical Bonds

It is very common in science classes to say that there is energy in chemical bonds, and that breaking these bonds releases energy. But this is exactly the opposite of what is true— energy is released when chemical bonds are *formed*!

Chemical bonds result from the electric force acting between positively charged atomic nuclei and negatively charged electrons, and this electric force gives rise to a potential energy associated with each possible arrangement of these charges. Just like boulders fall down cliffs rather than up them, charged particles also tend toward arrangements that have the least amount of potential energy. When atoms form bonds with each other, it is because the bonded arrangement is associated with a lower potential energy than the unbonded arrangement. When atoms are bonded, their potential energy is *lower* than it is when they are free. The term *bond energy* refers to the amount of energy input required to break a chemical bond such that the constituent atoms are free.

So, why is energy released when we burn something such as methane (in which four hydrogen atoms are bonded with one carbon atom)? This process happens in two steps.

First, we need an energy input (such as heat from a very hot object) to provide the atoms with sufficient energy that they escape their bonded state—this is like shaking a bucket of marbles vigorously enough so that some begin to escape the bucket. If we only excite them and do not provide them with other bonding options, they will reform the same bonds they had previous to heating and these bonds will have the same potential energy as they did before. The result will be no overall energy release. To get an energy release, the second thing we need is to provide the atoms with the opportunity to bond in a new arrangement that has a lower potential energy associated with it—this is the role oxygen plays in burning. When the carbon and hydrogen atoms in methane have the option to reform bonds with oxygen to form water (two hydrogen atoms and an oxygen atom) and carbon dioxide (two oxygen atoms and a carbon atom), they do so because these arrangements have a lower potential energy than the methane molecule did. Just like a boulder will fall until it reaches the lowest potential energy arrangement (lowest height above Earth) that it can, the free atoms will form bonds with the lowest potential energy arrangement possible. And just like the boulder-Earth system gains kinetic energy as the boulder falls to a lower potential energy arrangement, the newly formed molecules gain kinetic energy because the new arrangements have a lower potential energy.

Energy is released during the burning process not because the bonds of the burned substance release energy when they are broken, but because energy is released when the newly freed atoms reform chemical bonds with an even lower potential energy than they originally had. This decrease in potential energy must be accompanied by an increase in some other form of energy, such as the kinetic energy of the new molecules or the release of light.

Stored Energy

Often, scientists will say that molecules like methane (which is made of bonded hydrogen and carbon atoms) store energy. This is shorthand for saying that if we burn it (or use some other energy-releasing process), we can trigger a net release of energy when the atoms are broken apart and allowed to reform bonds with an even lower potential energy arrangement.

The term *store* can be interpreted as though energy were some material substance that is put into the atoms and kept there until we let it go again. Although this interpretation is not true, the idea of storing energy in chemical bonds is useful shorthand for scientists when talking about energy.

To form molecules that burn, such as methane (which is made of bonded carbon and hydrogen atoms) or glucose (which is made of bonded carbon, hydrogen, and oxygen molecules), we do indeed need an energy input. Although these molecules represent a lower potential energy arrangement than if the atoms were out there wandering free, the odds of finding free carbon, hydrogen, or oxygen atoms are extremely low. If these atoms are around one another, they will spontaneously bond in the lowest potential energy arrangement that they can, and water and carbon dioxide molecules are arrangements with a very low potential energy. Plants form glucose through photosynthesis (see Chapter 5), but the raw materials for glucose are not free carbon, hydrogen, and oxygen atoms; they are molecules of carbon dioxide and water! To form molecules of glucose (and oxygen molecules), which have a higher potential energy arrangement than the molecules of carbon dioxide and water, plants need to transfer energy to the molecules, and they get this energy from light. Photosynthesis is like pushing a boulder from the bottom to the top of a hill because it forms molecules with a higher potential energy arrangement than they were initially in; it requires an energy input to build the molecules, just like it does to get the boulder up the hill. In some sense, this is like keeping money in a bank account—it is simply there until you need it later. By forming glucose, plants have constructed a molecule with a high potential energy arrangement that can later react with oxygen to release energy again. This process functions a lot like our everyday process of storage, so scientists will often simply say that some molecules, such as the ones built in photosynthesis, store energy, without describing the whole energy process.

More broadly than chemical energy, it is common to refer to any form of potential energy as stored energy. You might hear people talk about the energy stored in a stretched bow and arrow or in a battery; both of these are cases in which a device maintains a high potential energy arrangement that can be used to release energy by allowing a transition to a lower potential energy arrangement. In general, this term refers to any time energy exists in a form where it is localized and easily accessible later. Energy can even be stored as the kinetic energy of a spinning flywheel or the thermal energy of hot water. Note that we say thermal energy can be thought of as stored; we did not use the word *heat*. This is because *heat* has its own very specific scientific meaning.

Heat

Like energy, the word *heat* was widely used in everyday language before it became a precisely defined scientific term. Children hear the word *heat* very early in their lives, and the everyday meaning of the word can be quite different from its precise scientific meaning. Students may hear that heat rises when discussing hot air balloons, they may be asked to turn the heat up on a cold day, or they may make a rash choice in the heat of the moment. In everyday language, *heat* can mean a variety of things, but the term is very precisely defined in science.

As a scientific term, *heat* describes the transfer process by which energy moves between regions of different temperature. Heat is related to temperature, but these are distinct scientific concepts. Temperature is a measure of the average kinetic energy associated with the random movements of particles in an object. The higher the temperature of an object, the faster its particles vibrate or move. The energy form associated with this random motion of particles is called thermal energy (not heat energy). All objects have thermal energy because all molecules are in motion, and the faster the particles in a system are moving, the more thermal energy it has. Thermal energy is essentially a measure of the total energy in an object or system arising from its random particle motion. Thus, a system can have more thermal energy by having either faster-moving particles or more particles. So, temperature measures the average kinetic energy of particles, thermal energy measures the total energy of a system arising from random particle movements, and heat occurs when these moving or vibrating particles transfer energy to their surroundings.

To keep the conceptual boundaries between heat and thermal energy clear, a good rule of thumb is to use the word *heat* only as a verb. Saying "heat the water" implies that you transfer energy into the water to increase its temperature. On the other hand, the phrase "hot water has a lot of heat" blurs the line between heat and thermal energy. When systems are heated and their temperature increases, this change in energy is an indication that the thermal energy of the system has increased as well.

There are three mechanisms by which energy can transfer to or from systems via heat: conduction (in which fast-moving particles collide with slower-moving particles to transfer energy), convection (in which fast-moving particles move from one region to another), and radiation (in which vibrating particles emit energy as electromagnetic waves). In each of these processes, energy moves from place to place. Scientists originally thought of heat transfer as involving a very special fluid called caloric, but we now understand that there is no such fluid being transferred. Yet, the word *flow* has remained part of the scientific vocabulary for energy.

Energy Flow

When energy transfers between systems and spreads out within the largest region it can, it acts a lot like a fluid. Scientists use the term *energy flow* to describe energy transfer between systems because there are parallels between the transfer of energy and fluid or air flow. If you used a funnel to pour 2 liters of water into a 2-liter bottle, you'd be pretty surprised if the bottle only filled up halfway—that is, you'd be surprised that during this transfer of water, the amount that left your cup was not the amount that entered the bottle. If you noticed this was, in fact, the case, you'd start to look for leaks rather than think that the water just spontaneously disappeared in the process of being transferred. Fluid flow is a useful analogy for energy transfers between systems because most of us have developed an intuition that fluid that flows from someplace must end up someplace else, and because we understand that fluids tend to spread out to fill as much of their container as they can. Energy flow, however, is different from fluid flow because energy is not a material substance—its flows are typically tracked as changes in energy forms across systems.

Although using the term *flow* to describe energy transfers isn't perfect, it does have the major benefit of helping students use their intuition about fluids to bolster their sense that energy missing from one place must show up someplace else. That is, energy is conserved.

Building Energy Language Over Time

We learn science words just like we learn every other word: with time and practice in a variety of contexts. Furthermore, we learn words when we need them and distinguish among multiple meanings of the same word when it is time to do so. We would not expect kindergarteners to describe their relationships with classmates as amicable because this word implies a rather adultlike understanding of interpersonal relationships. Likewise, we would expect kindergarteners to think that a cookie is a delicious treat; we would not expect them to realize that a cookie is also a packet of data sent by an internet server to be stored by a user's web browser.

As they learn the energy concept, students begin to use new words and distinguish between everyday and scientific meanings of familiar words. In this process, new words should be introduced when they are necessary and the multiple meanings of words should be distinguished only when there is a reason to do so. Students will have heard and used words such as *energy, heat,* and *work* in their everyday lives before they encounter them in a scientific setting; using the nonscientific meaning of these words is okay until it is not. Investigations of phenomena should precede the introduction of new terms and meanings, and these new language tools should be introduced when they are useful for clarifying concepts.

This does not mean that teachers are off the hook for using appropriate energy language throughout. Nobody defines the word *chair* for us when we are young, but as we hear adults use the term in a conceptually consistent way, we develop an understanding of the term and an ability to use it in consistent ways that mirror how the adults use it. If we are sloppy with energy language in the science classroom, this can make it difficult for students to discern the conceptual delineations of a term. For example, we needn't define the term *heat* for young students or insist that they use its scientific meaning, but by using the word in ways that are consistent with heat as a process (rather than, say, a property of an object), students are in a much better position to use the word appropriately themselves later on.

Just as we should use scientific terms correctly even though we may not expect our students to do so, we need to be careful using language shortcuts before they have been earned by students in our classes. For example, using the term *energy flow* too soon may promote the idea that energy is some sort of physical fluid.

So, what do we say and when do we say it? The *NGSS* are a good guide.

Talking About Energy in Elementary School

The fourth-grade performance expectations for energy in the *NGSS* emphasize energy transfer and conversion (i.e., transformation). By design, the performance expectations focus on energy transfer processes that are easy for students to observe. These transfer processes include heat, sound, light, and electricity; students should not be expected to precisely define these terms, but they should be able to use them in the context of discussing the role of energy transfers in natural phenomena and designed devices.

The major emphasis at the elementary level is on building students' intuition for the role of energy in driving processes. There is no attempt at this level to define *energy*—only to associate the term with things such as motion, sound, light, heat, and electricity. Rather than distinguishing among energy forms at this level, the emphasis is on seeing how energy transfers and motion are really manifestations of the same thing. Thus, at the elementary school level, language should reflect unifying energy-related phenomena under the same conceptual umbrella, by recognizing that something called energy can be transferred between objects and systems in a variety of processes.

Terms such as *potential energy, kinetic energy, conservation,* and *work* are inappropriate and unnecessary for energy instruction at the elementary level. On the other hand, introducing and consistently using terms such as *energy transfer* and *energy conversion/transformation* are critical to helping students understand that energy is associated with certain observable characteristics of systems and that energy can transfer from place to place in a variety of ways. It's too early at this age to use shortcut words such as *flow;* students should instead practice using the language of transfer and conversion.

Talking About Energy in Middle School

The middle school performance expectations for energy in the *NGSS* focus on helping students develop a more sophisticated understanding of energy transfer and transformations. By advocating the introduction of ideas such as potential energy in middle school, the *NGSS* help students build toward a deeper understanding of energy as a conserved quantity (though this idea is not yet introduced).

In middle school, students should begin to use language that reflects a more nuanced understanding of energy. While elementary students focus on relating easily observable phenomena (i.e., sound, light, heat, electricity, and motion) around the unifying idea of energy, middle school students are developing an ability to think about energy in a more abstract manner. Rather than focusing on the overt phenomena, middle school students begin to understand that their observations are merely *indicators* of different forms of the same underlying quantity called energy.

To reflect a more sophisticated understanding of the difference between their observations and the unobservable, unified quantity called energy, students should begin to use terms describing various forms of energy, such as *kinetic energy, thermal energy,* and *gravitational energy,* and to describe how each is related to what they observe. For example, while elementary students should associate the idea of energy with motion, middle school students should begin to call this energy kinetic energy. Calling the energy of motion kinetic energy corresponds with an increased emphasis on uncovering its precise relationship with mass and speed. That is, the term is not simply introduced because it is the next more sophisticated word, but because the term has a precise scientific meaning that students in elementary school are not yet ready to describe.

Just as introduction of the term *kinetic energy* at middle school corresponds with an increased need and ability to specifically describe it, students at the middle school level will begin to use other terms that reflect concepts that are not introduced in elementary school. For example, they should begin to use the term *potential energy* to describe the energy associated with objects that interact at a distance and should be careful to associate this term with systems of objects rather than individual objects.

As students analyze phenomena from an energy perspective, they should describe changes using terms such as *energy transformation* (between energy forms) and *energy transfer* (between systems) and connect these processes to the evidence that they are occurring. By consistently using the language of transformation and transfer, students gain practice identifying various forms of energy and connecting observations to changes in energy within and between systems. For example, students in middle school become ready to identify heat as an energy transfer process rather than an energy form, but initially using a term such as *thermal energy transfer* to describe heat exchange processes can help students recognize that the process of heating and cooling requires the transfer of energy between systems at different temperatures.

As students at the middle school level expand their energy vocabulary, the emphasis should be on introducing words that help them understand and explain the processes of energy transformation and transfer. As their familiarity with the ideas of transfer and transformation grows, students begin to earn linguistic shortcuts such as *flow,* but such terms should only be introduced after students have had sufficient practice identifying and describing phenomena using the language of transformation and transfer. By regularly using language focused on energy forms, transformations, and transfers, students are in a better position to use these terms consistently throughout the course of the year as they study phenomena from the physical, life, and Earth sciences and to use language shortcuts in appropriate ways.

Talking About Energy in High School

The high school performance expectations for energy in the *NGSS* focus on helping students understand and use energy as a quantitatively conserved quantity. Students begin to calculate energy changes within systems and transfers between them. As students begin to transition into discipline-specific science courses in high school, it becomes even more important for teachers to be aware of how discipline-specific energy language can make it more difficult for students to see connections among the role of energy in phenomena across disciplines.

Different science disciplines talk about energy in different ways. Terms such as *energy flow* and *energy use* are more commonly used in a life science context than in a physics context, whereas terms such as *conservation* are more commonly used in a physics context. The simple fact is that certain dimensions of energy are less critical in some contexts than others, but if we hope to help students connect ideas across disciplines, it is important to use terminology that puts students in a strong position to see the overlap in energy ideas across disciplines. Energy terminology may vary among disciplinary contexts, but the rules of energy remain the same.

In physical science classes, students commonly calculate energy transfer into and out of systems via processes such as work and heat. Earth science classes commonly identify sources of energy that drive cycles. Life science classes often discuss energy flows. In the context of energy, words such as *work, heat, source, drive,* and *flow* all involve energy transfer, but without making this connection explicit, such disparate terminology can contribute to student confusion and difficulty connecting ideas across disciplines. Thus, it is especially important for disciplinary high school teachers to define such discipline-specific terms in terms of the Five Big Ideas that make energy crosscutting.

By connecting discipline-specific terms to energy forms, transformations, transfers, conservation, and dissipation, teachers put students in a better position to connect ideas and use what they learn in one discipline to inform their learning in another. If, for example, Earth science students connect the idea of source to the idea of transfer between systems, then they may be more likely to go beyond identifying the Sun as the source of energy for the water cycle and realize that any transfer into a system must equal the energy change in the system plus the energy transferred from it. Such a realization might spur questions about how the balance of different states of water on Earth affects global warming. Likewise, using the language of flow in a physics class exploring the "work-in, work-out" balance in simple machines may help students connect their exploration of simple machines to more meaningful and complex systems, such as living organisms, that they encounter in other disciplinary contexts.

Summary

Language is a critical tool in science for formulating and communicating ideas. To describe the behavior of energy, scientists have borrowed many words from everyday contexts. The overlap of some energy-related words in scientific and everyday contexts can be both a blessing and a curse—it can help connect energy learning with students' intuitive ideas, but it can also confuse the precise scientific meaning of energy terms. When teaching and learning about energy, it is especially important to use language carefully.

To use language carefully in energy instruction, consider which terms are important to define in scientific contexts, which ones can be used colloquially (but consistently) without stressing the scientific meaning, and which ones should be avoided entirely as students are building their understanding within each grade band. The *NGSS* are very intentional about how they include energy-related terms across the science disciplines and during grades K–12, and they serve as a good guide for when and how to use particular terms related to energy.

By identifying energy as a crosscutting concept, *A Framework for K–12 Science Education* (NRC 2012) and the *NGSS* call on teachers to help students understand how a consistent set of energy ideas is applicable across disciplines. To do this, students need language that is unified around a small set of core ideas about energy. The Five Big Ideas about energy help do this, but even these Five Big Ideas use terms that have different meanings in scientific and everyday contexts. Thus, the call to build a consistent set of energy ideas that are crosscutting also involves a call for all of us to be more careful with our language as we help students build and discuss their ideas about energy over time.

References

Coopersmith, J. 2010. *Energy, the subtle concept: The discovery of Feynman's blocks from Leibniz to Einstein*. New York: Oxford University Press.

Hecht, E. 2003. An historico-critical account of potential energy: Is PE really real? *The Physics Teacher* 41 (8): 486.

Lemke, J. L. 1990. *Talking science: language, learning, and values*. Norwood, NJ: Ablex Publishing.

National Research Council (NRC). 2012. *A framework for K–12 science education: Practices, crosscutting concepts, and core ideas*. Washington, DC: National Academies Press.

NGSS Lead States. 2013. *Next Generation Science Standards: For states, by states*. Washington, DC: National Academies Press. *www.nextgenscience.org/next-generation-science-standards*.

Quinn, H. 2014. A physicist's musings on teaching about energy. In *Teaching and learning of energy in K–12 education*, eds. R. F. Chen, A. Eisenkraft, D. Fortus, J. S. Krajcik, K. Neumann, J. C. Nordine, and A. Scheff, 15–36. New York: Springer.

Vygotsky, L. S. 1978. *Mind in society: The development of higher psychological processes*. Cambridge, MA: Harvard University Press.

CHAPTER 5
ENERGY IN PHOTOSYNTHESIS AND CELLULAR RESPIRATION

JOSEPH KRAJCIK AND JEFFREY NORDINE

Where do you get your energy to do all the things you need to do? Why does eating food provide you with energy? Let's look at a common phenomenon to help you understand where you get energy from: the burning of wood or straw (biomass) with oxygen (O_2). In this phenomenon, chemical energy in the biomass-oxygen system is converted to light energy and thermal energy, and molecules of carbon dioxide (CO_2) and water (H_2O) are formed. What does this have to do with where you get your energy?

Let's revisit the cheese puff lab discussed in Chapter 3. When we place a cheese puff on the head of a pin and burn it, thermal energy and light are released. We can feel the thermal energy and we can see the light. The atoms in the molecules that make up the cheese puff (e.g., starch, fat, and protein molecules), plus the atoms from the oxygen molecules in the air, rearrange to produce carbon dioxide and water. The rearrangement releases light and thermal energy. Chemical energy in the cheese puff–oxygen system is transformed to thermal energy associated with the fast-moving molecules of carbon dioxide and water that are formed when the cheese puff burns, and a bit of light is produced as well. You can use the release of the thermal energy to do things such as heating water or turning a wheel. As the cheese puff lab shows, we can use the thermal energy of the fast-moving molecules of carbon dioxide and water to transfer energy to liquid water. It is incredible how much thermal energy a little cheese puff can give off! There is a transformation (or conversion) of the energy stored in the cheese puff–oxygen system to light and thermal energy of the fast-moving molecules of the carbon dioxide and water products.

Notice that, to start the burning reaction, an energy input is needed—we must heat the cheese puff a bit to get it to start reacting with oxygen in the air. Why is this energy needed? As atoms and molecules move and vibrate faster, they eventually have enough kinetic energy to escape the bonds that hold them together. When the products of the burning reaction are formed, they form new bonds in an arrangement that has a lower potential energy than they had in the first place (when they were reactants). The difference in the high potential energy associated with the reactants and the low potential energy associated with the products is now manifested as additional kinetic energy in the products. Just like a falling object moves faster as gravitational potential energy decreases, the molecules that are the products of burning start moving faster as chemical potential energy decreases. This additional kinetic energy continues to drive the reaction, and it also transfers thermal energy to the surroundings as the products collide with molecules that were not involved in the reaction. If the products were not at a lower potential energy than the reactants, the reaction would not be able to continue on its own.

Energy is released when we burn food such as cheese puffs, but where did this released energy come from in the first place? How is the burning of biomass or cheese puffs related to how you get energy when you eat? In a process called *photosynthesis,* plants use energy from the Sun, as well as carbon dioxide and water, to make glucose (a type of sugar with the chemical formula $C_6H_{12}O_6$) and oxygen. In a process called *cellular respiration,* our bodies use these glucose molecules and oxygen to make carbon dioxide and water and release the energy necessary for driving life processes.

In this chapter, we will explore in greater depth the ideas presented in the example above and dive into two of the most important phenomena that occur in our world that allow life to exist on Earth—photosynthesis and cellular respiration. First, we will explore how plants make use of energy from the Sun, carbon dioxide, and water to produce sugar and oxygen in the process of photosynthesis. Then, we will explore how living systems use sugar and oxygen to release the energy necessary to carry out important life functions such as moving and staying warm. Although photosynthesis and respiration are typically considered part of the biology curriculum, the ability to explain photosynthesis requires interdisciplinary understandings of several science disciplines. A deep understanding of photosynthesis includes knowledge of how matter interacts and key ideas about energy—particularly energy transformation (Big Idea 2), transfer (Big Idea 3), conservation (Big Idea 4), and dissipation (Big Idea 5) (see Chapters 1 and 2 for a description of the Five Big Ideas).

In this chapter we will

- provide an overview of why photosynthesis is so important, what students should know about it, how it relates to activities such as the cheese puff lab, and how solar energy is transferred from the Sun to drive photosynthesis;

- take a deeper dive into photosynthesis as a chemical process that transforms matter and energy;

- discuss cellular respiration as a chemical process by which plants and animals use food by transforming matter and energy;

- consider how students should build ideas about photosynthesis and cellular respiration over time; and

- present teaching approaches for helping students gather evidence that supports their understanding of photosynthesis and cellular respiration.

Why Is Photosynthesis So Important for Our Life on Earth?

The photosynthesis process, in which plants use light from the Sun to produce their own food from carbon dioxide and water, generates products that support all animal life on Earth and that ultimately allow our society to function. This incredible process provides access to energy for all plant and animal life, including humans, to live and maintain warmth, and it also makes it possible for humans to have the residential, commercial, and industrial systems on which we rely so heavily. It is wondrous to think about how our lives depend so much on plants and energy from the Sun!

The fuels we need to power our societies come mainly from the process of photosynthesis in plants. The United States obtains about 84% of its total energy from various forms of fossil fuels (i.e., oil, coal, and natural gas) (National Academy of Sciences 2015). These fossil fuels were derived from materials originally formed by plants, which capture solar energy to drive the process of photosynthesis. Humans burn these fossil fuels to release the energy necessary for operating our electric appliances, heating our houses, driving our cars, and operating many other devices important to our daily lives. Many countries are now exploring the capture of solar energy to generate electricity, which can be used for lighting, heating, powering cars, and other purposes. Although transportation still chiefly depends on fossil fuels, the use of vehicles powered by electricity and alternative fuels (e.g., solar, hydrogen, and biofuel) is increasing (National Academy of Sciences 2015).

How Do Plants Capture Energy?

The energy available for life on Earth comes almost exclusively from the Sun.[1] In this section, we focus on how plants capture energy from the Sun. Sunlight is the predominant source of energy that we use on Earth, and it provides a surprisingly large amount of energy. We all have experienced this solar energy when we feel warmer as sunlight shines down on us. When it is overcast or we stand under a shade tree, we feel cooler. If we could convert all the solar energy that strikes a single square meter of Earth's surface into electrical energy (we can't!), it would provide enough energy to power a typical stereo system, laptop, and small television set (ConsumerReports.org 2014; Lindsey 2009). That is a tremendous amount of energy available in an area about the size of card table.

Many scientists and engineers have devoted their careers to the challenge of developing new ways to capture and use energy from the Sun. New solutions to this challenge could address many problems facing society, such as the increasing concentration

---

1 Some energy to support life and society comes from thermal energy from hot materials deep inside Earth and energy released in nuclear fission (both natural and human-initiated) of large atoms found naturally in Earth, such as uranium. But plants are fundamental for supplying energy to our bodies.

of carbon dioxide in the atmosphere. This is an active area of research in many countries throughout the world.

A Framework for K–12 Science Education (NRC 2012) emphasizes that even young students need to develop knowledge that all living organisms—plants and animals—require a source of food to perform the functions required to live and grow. All organisms use food to grow and repair tissues and provide energy for basic life functions. In addition to carrying out basic life functions, most animals need access to energy to hunt, reproduce, and maintain body temperature. Although virtually all energy for life ultimately comes from the Sun, both plants and animals must use a complex series of chemical reactions to access this energy. The complete series of chemical reactions required to support life are extraordinarily complex and beyond the scope of the *Framework* and the *Next Generation Science Standards* (*NGSS;* NGSS Lead States 2013).

Although plants and animals both need food, they obtain it in very different ways. Young learners need to realize that plants can produce their own food, which they use for growth, repair, and various life functions, if they have access to the appropriate factors from the environment, in particular light, water, and air (i.e., carbon dioxide). Later on, students must come to understand that plants and animals obtain and use food through specific cellular and molecular processes that support life functions. Plants make their own food through photosynthesis, but animals digest food that they have obtained from another organism—either plants or animals that have consumed plant matter and changed it into other molecules through chemical reactions. Moreover, the *Framework* emphasizes that students need to focus on the key role of specific molecules involved in these processes (in particular, oxygen, carbon-containing molecules such as glucose and carbon dioxide, and water). The *Framework* states that an atomic-molecular perspective on photosynthesis should be introduced in middle school and become more sophisticated in high school. The grade band endpoint for eighth grade for LS1.C states, "Plants, algae (including phytoplankton), and many microorganisms use the energy from light to make sugars (food) from carbon dioxide from the atmosphere and water through the process of photosynthesis, which also releases oxygen" (NRC 2012, p. 148).

The *Framework* also stresses how glucose in animals and plants can recombine with atoms from nutrients to form new larger structures (e.g., macromolecules such as starches, fats, proteins, and DNA), which, in turn, are used by organisms to fulfill important life functions and growth. Three disciplinary core ideas (DCIs) are particularly important to our discussion: PS3.D: Energy in Chemical Processes and Everyday Life; LS1.C: Organization for Matter and Energy Flow in Organisms; and LS2.B: Cycles of Matter and Energy Transfer in Ecosystems.

The process of photosynthesis is a series of chemical reactions that captures light energy and stores it as chemical energy by converting carbon dioxide and water into glucose (sugar) plus released oxygen. The arrangement of atoms in sugar and oxygen is at a higher potential energy state than the arrangement of atoms in carbon dioxide and water. The sugar molecules that are formed contain carbon, hydrogen, and oxygen, and these molecules are used to make amino acids and other carbon-based molecules that can be assembled into the larger macromolecules (such as proteins or DNA) needed for various life functions. The arrangement of the atoms in these macromolecules is at a very high potential energy state. During cellular respiration, organisms use oxygen and food (sugars) and produce water and carbon dioxide with the transformation of energy for various life functions.

How Do We Get Energy From Food? A Deeper Dive Into the Cheese Puff Lab

How is all of this related to the burning cheese puff that started this chapter? Where does energy in the cheese puff come from? The arrangement of atoms in the cheese puff–oxygen system is at a higher potential energy state than the arrangement of atoms in carbon dioxide and water that is produced during burning, and this difference in potential energy states makes energy conversion and transfer processes possible.

Let's think of the water behind a dam in a hydroelectric plant as an analogy to help us understand what is going on when a cheese puff burns. The water behind the dam is in a higher potential energy state and moves to a lower potential energy state as it flows through the dam to a lower elevation. As it flows downhill, water begins to move rapidly and its movement can be used to drive a turbine to produce electricity, which can then be used to power devices such as a lightbulb, a motor, or a heating coil. As a hydroelectric plant operates, there is a conversion of potential energy to kinetic energy, then the kinetic energy of the water transfers to kinetic energy of the turning turbine, and, finally, the turbine produces electricity, which transfers energy to our devices. In this process, energy is converted from one form to another (Big Idea 2) and then transferred from one object to another (Big Idea 3). A hydroelectric plant does not produce energy as it operates (Big Idea 4)—the energy carried by the electricity was originally potential energy associated with the water behind the dam. But the energy carried by the electricity is less than the original potential energy. Energy is also not destroyed in this process of energy conversions and transfers in a hydroelectric plant, but much of the energy does become thermal energy that spreads out and becomes difficult to use (Big Idea 5) for powering our devices.

Just like the water behind a dam, the cheese puff–oxygen system contains potential energy. When molecules from the cheese puff react with oxygen, potential energy is transformed to light and thermal energy associated with the fast-moving carbon dioxide and water molecules released during the chemical reaction (Big Idea 2). We can use the thermal energy released during the burning of the cheese puff to heat water as the fast-moving carbon dioxide and water molecules collide with molecules around them (Big Idea 3). Although we may want these fast-moving molecules to transfer all of their energy to our container of water, they will simply collide with whatever they happen to hit, and there are plenty of molecules to hit in the surrounding air. Air is a mixture of molecules of nitrogen, oxygen, carbon dioxide, and trace amounts of lots of other types of molecules. When these molecules speed up, they may just happen to hit the water container and transfer energy to it, but many of them simply spread out into the environment and collide with other molecules in the air (Big Idea 5), thus increasing the temperature of the air instead of the temperature of our water. Just like the hydroelectric plant, the burning cheese puff does not produce energy (Big Idea 4). Rather, the process of burning a cheese puff to heat water involves a conversion from potential to kinetic energy and a transfer of the kinetic energy of one object to the kinetic energy of another object (i.e., the heating of the water); in this process, not all of the energy ends up where we might like it to end up.

As we burn the cheese puff in the cheese puff lab, it acts as a fuel. Sugar, gasoline, wood, coal, and dry grass are all examples of fuel because all can burn (combust) to release energy. All combustible fuels can react with oxygen and release energy, although burning some fuels releases more energy than others. All food and combustible fuels contain complex molecules that have carbon as their base, and these complex molecules are abundant on Earth because of the process of photosynthesis.

When we burn a fuel such as sugar, the potential energy of the carbon dioxide and water (the products of burning) is much less than that of sugar and oxygen (which react during burning). When plants photosynthesize, they make sugar and oxygen from carbon dioxide and water, and this process *increases* potential energy! This is like water flowing uphill. To accomplish this, energy needs to transfer into the carbon dioxide–water system to drive the formation of sugar and oxygen, but in a very controlled manner. Plants achieve this incredible feat through the process of photosynthesis, which uses solar energy to increase the potential energy arrangement of atoms in molecules. This process is carried out in the chloroplasts of plants, and it is quite complex. Plants contain very sophisticated chemical machinery that can capture (transfer) solar energy (i.e., radiation) and use it to break bonds between atoms and form new bonds with a higher potential energy arrangement.

It is important to realize that a full explanation of the photosynthesis process goes beyond what is presented in the *Framework* and the *NGSS*. Although it is not necessary for students to learn all the complex details, one thing all students must learn about photosynthesis is that the process cannot happen without an energy input via light. In the next section, we will explore in more depth how light works to transfer energy over a distance of 93 million miles from the Sun to drive the construction of complex molecules in plants on Earth.

What Is Solar Energy?

Light from the Sun travels outward in all directions, and a tiny fraction of this light reaches Earth. When it hits Earth, some is reflected back into space and some is absorbed. When light is absorbed, its energy is transferred to the object that absorbed it. When the ground absorbs light, it warms up; when plants absorb light, they can use its energy to make sugars and release oxygen.

Light is a form of electromagnetic radiation. The word *electromagnetic* means that it is made up of electric and magnetic fields, and the word *radiation* means that it travels outward from a source. Electromagnetic radiation travels outward from a source as waves of electric and magnetic fields in space. Electromagnetic radiation from the Sun is also referred to as *sunlight, solar energy,* or *solar radiation.* All of these terms refer to energy transferred from the Sun in the form of electromagnetic waves. Unlike thermal energy or sound, electromagnetic energy does not require a medium and can travel through empty space. Just like waves on a rope, electromagnetic waves can oscillate quickly or slowly, but they all travel through space at an extremely high speed (*much* faster than waves on a rope).[2,3] These ideas about electromagnetic radiation are part of two other DCIs, PS4.A: Wave Properties and PS4.B: Electromagnetic Radiation.

Most of the electromagnetic radiation that comes from the Sun oscillates in a relatively small range of frequencies, and our eyes are sensitive to a portion of the frequencies in this range, which we call *visible light* (see Figure 5.1, which shows the entire electromagnetic spectrum). If the frequency is a little too low for our eyes to see, we call it *infrared* (below red); if it is a little too high, we call it *ultraviolet* (above violet). Solar radiation is mostly made up of infrared, visible, and ultraviolet light.

Figure 5.1. The electromagnetic spectrum

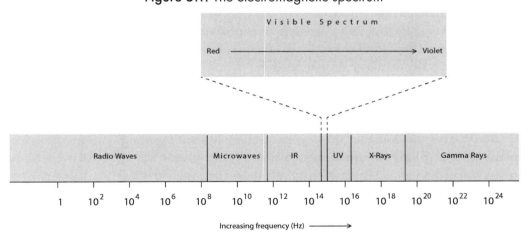

Note: The visible spectrum, which is just a tiny fraction of the whole electromagnetic spectrum, contains the ROYGBIV (red, orange, yellow, green, blue, indigo, violet) colors that we see. IR = infrared; UV = ultraviolet.

---

2 You might have made waves on a rope as a child. Two people hold either end of a rope, and one person shakes his or her arm to send waves along the rope, which travel to the other person. If one person shakes his or her arm quickly or slowly, the other person will feel the rope vibrate back and forth quickly or slowly as well.

3 Regardless of its frequency (how quickly its electric and magnetic fields oscillate), all electromagnetic radiation travels at about 300,000,000 meters per second (or 186,000 miles per second)—this is the speed of light. At this speed, light travels the 93 million miles from the Sun in about 8 minutes.

Plants contain specialized molecules and structures that allow them to absorb some of the solar radiation that reaches them. Green plants contain chlorophyll molecules,[4] which are light-absorbing pigments. So, which color of light from the Sun might you expect is most useful for photosynthesis? It's tempting to think that it's green, but photosynthesizing plants look green because they *reflect* green light (our eyes detect this reflected light and we see green). It turns out that the chlorophyll molecules in green plants mostly absorb light in the blue and violet parts of the visible light spectrum and also the deep red part. You can try a little experiment: Go into a very dark room and shine a red light on a plant. What color do you see? The plant should look black, because most of the red light is absorbed and not reflected back for our eyes to see.

When the chlorophyll molecules in plants[5] absorb solar energy, this energy drives the complex chemical machinery that carries out the "uphill" process of constructing sugar and oxygen molecules (higher potential energy) from carbon dioxide and water molecules (lower potential energy). If it weren't for photosynthesis, the solar energy transferred to the Earth would simply heat it or be reflected back to space. But because chlorophyll molecules can absorb solar energy and use it to drive the formation of complex molecules, life on Earth is possible.

A Deeper Dive Into Photosynthesis

All living systems, including humans, require an energy input to grow and reproduce. Animals obtain energy from eating food, and although some humans and other animals eat principally other animals, most animals eat plants. Plants also need food to grow and reproduce. But, unlike animals, plants build their own food through photosynthesis. The word *photosynthesis* contains clues to its meaning, and you should point these clues out to your students. The prefix *photo* comes from a Greek word meaning "light," and the root of the word *synthesis* is from the Greek word meaning "to put together." *Photosynthesis* literally means "using light to put together." Plants use light to put together food from the raw materials of carbon dioxide and water found in the environment. When animals eat plants, they consume the food that plants have made, and energy is transferred from the plant system to the animal system.

Through photosynthesis, plants use solar energy (i.e., light) to construct glucose (a type of sugar) using water and carbon dioxide as raw materials. Figure 5.2 shows the overall chemical reaction for photosynthesis. As we will describe later, photosynthesis is actually a series of chemical reactions, but the overall result is the same as if it were one chemical

---

4 The discussion of chlorophyll molecules in this and the next section goes beyond what is in the *Framework* and the *NGSS*.

5 Incredibly, chlorophyll molecules are not just found in plants—they can also be found in many microorganisms and even some prokaryotic cells. These organisms can also carry out photosynthesis. Phytoplanktons in the ocean are the primary source of food for other organisms that live in water systems, and these organisms account for roughly half of the photosynthetic activity on Earth (Carlowicz 2009).

Figure 5.2. The overall chemical reaction for photosynthesis

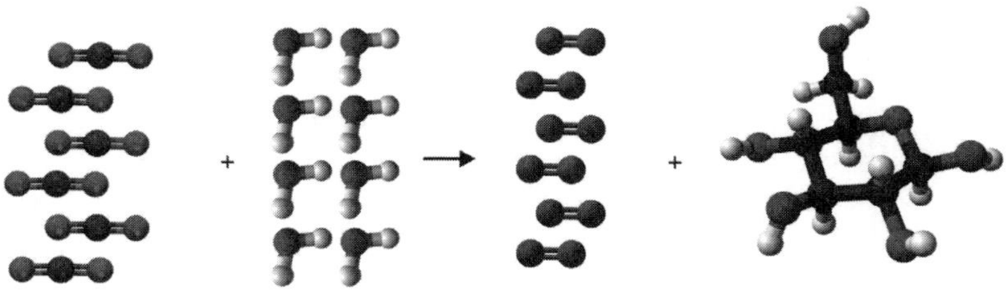

| 6 Carbon Dioxide | + | 6 Water | → | 6 Oxygen | + | 1 Glucose |
| $6CO_2$ | | $6H_2O$ | | $6O_2$ | | $1C_6H_{12}O_6$ |

Source: Adapted from Krajcik and Drago 2013. Used with permission.

reaction that rearranges the atoms from carbon dioxide and water into molecules of glucose and oxygen. Because the products of photosynthesis have more potential energy than the reactants (raw materials), the photosynthesis process can occur only because plants absorb energy from the Sun—that is, because energy is transferred to them from their surroundings. The glucose (sugar) that is produced in photosynthesis is used in various forms by all of the organisms that live on this planet.

Figure 5.3 (p. 88) is a representation of the photosynthesis process showing the importance of the chlorophyll absorbing light. Notice how the root system is important in taking in water from the ground and how the leaves have special structures for taking in carbon dioxide from the atmosphere and releasing oxygen. It seems a little odd to think that oxygen, so necessary for life on Earth, is really just a by-product of photosynthesis! The photosynthesis process is essential in recycling the carbon dioxide and producing sugar, but the production of oxygen is just as vital.

So far, we have said that light is critical in driving the process of photosynthesis, but just how does the energy in sunlight get captured? Earlier in this section, we discussed light in terms of (electromagnetic) waves, but light also behaves as a particle called a *photon*.[6] You can think of a photon as carrying a specific amount of energy or a bundle of energy. Whether or not a molecule can absorb a photon with a particular energy depends on the structure of the molecule or, more specifically, the electrons that surround the nuclei of the atoms that make up the molecule. When a photon interacts with (i.e., strikes) a molecule capable of absorbing it, the energy of the photon is absorbed by some of the electrons, and these electrons are moved to a higher potential energy state. This is what happens in plants.

A plant has specialized chlorophyll molecules, which are light-absorbing pigments that capture photons of electromagnetic radiation containing a specific amount of energy that is due to oscillation at a specific frequency. Because energy is conserved (never destroyed), the

---

6 If this doesn't make sense to you, join the club! This is a quantum mechanical phenomenon that makes very little intuitive sense, but it is the way that nature behaves.

Figure 5.3. A diagrammatic view of the photosynthesis process

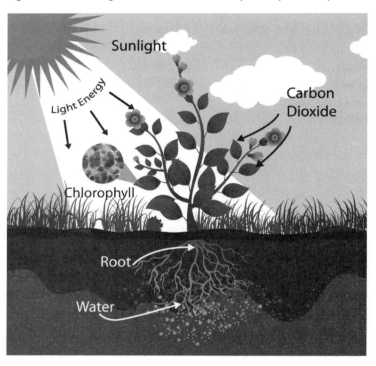

energy transferred from sunlight has to go somewhere. The solar energy transferred to the molecule that absorbs it is manifested as increased potential energy as a single electron is "kicked up" to a higher potential energy state. Think of this as a ball being moved from the ground to the top of a table. When the ball is on the table, the ball-Earth system has greater potential energy than when the ball is on the ground, and this increase in potential energy is due to the attractive gravitational force between the ball and Earth. Similarly, when an electron moves to a higher potential energy state, this higher energy is due to the attractive force between the negatively charged electron and the positively charged nucleus it orbits. Just as a ball will roll off a table given enough time, the electron in a higher potential energy state will eventually find its way back to a lower potential energy state. There are two possible ways that this decrease in energy can happen within a chlorophyll molecule: (1) the molecule can revert to the lower potential energy state by releasing energy in the form of electromagnetic radiation (emit a photon) as the electron falls back to a lower state; or (2) the molecule can react with another molecule by giving up an electron to another molecule, thereby transferring energy to the second molecule. Option 2 happens in photosynthesis.

The chlorophyll molecules in plants absorb energy that is transferred from the Sun via light, and this absorbed energy is used to build two molecules that are at a higher energy state (i.e., have a higher potential energy). Plants then use these high-energy molecules to rearrange the atoms in carbon dioxide and water into glucose and oxygen.

On its own, the glucose produced in photosynthesis is useless. That is, the high potential energy associated with the arrangement of atoms in glucose is inaccessible unless it reacts with oxygen so that atoms can arrange themselves into a lower potential energy state. But this rearrangement won't just happen on its own! Think about it—there are oxygen molecules hitting the cheese puff all the time, but do you ever worry about your bag of cheese puffs spontaneously catching on fire? Glucose and oxygen molecules are quite stable, even though they are at a relatively high potential energy arrangement. To get glucose to react with oxygen, we can heat it (by quite a bit!) so that it starts burning. As it burns, we see a rapid release of energy. To get food molecules and oxygen to react such that there is a steady, controlled release of energy is another incredible feat of life. This feat is called cellular respiration.

Cellular Respiration

All living organisms (including plants and animals) use the energy released by the reaction of sugar and oxygen to carry out various life functions such as reproduction, growth, and tissue repair. It takes a tremendous amount of energy just to remain alive. A typical human requires an average energy input of about 2,000 calories per day to maintain basic life functions. (This figure varies somewhat by person and is called our *basal metabolic rate*.) On average, this is enough energy to keep a 55-inch HDTV on all day and all night (based on energy use of the Sony W800B Premium LED HDTV). To obtain this energy, we must not only eat food but also breathe in oxygen, because the molecules in food will not spontaneously rearrange into a lower potential energy state on their own.

Once in our bodies, molecules in the food we eat are involved in a complex series of chemical reactions as they are digested and converted into a form useful by our cells to carry out cellular respiration. Like photosynthesis, cellular respiration can be simplified by looking just at its inputs and outputs as if it were a single chemical reaction (see Figure 5.4, p. 90). Notice how the overall chemical reaction for cellular respiration is the reverse of the overall reaction for photosynthesis. Photosynthesis absorbs energy to build complex molecules, whereas cellular respiration releases energy (i.e., transfers it to other systems) as it dismantles these molecules to form a lower potential energy arrangement.

Like animals, plants must also go through the process of cellular respiration to remain alive. Many people are confused by the idea that plants also go through cellular respiration to provide the energy for basic life functions. But rest assured, plant cells don't run on light—they run on food. They just use light to provide them with energy necessary to make their own food.

Figure 5.4. The overall chemical reaction for cellular respiration

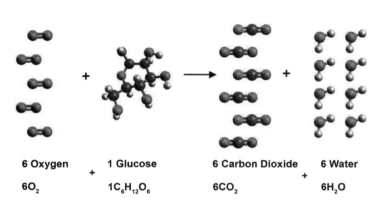

6 Oxygen	1 Glucose	6 Carbon Dioxide	6 Water
+		+	
$6O_2$	$1C_6H_{12}O_6$	$6CO_2$	$6H_2O$

Source: Adapted from Krajcik and Drago 2013. Used with permission.

Learning About Photosynthesis and Respiration Over Time

The *Framework* presents a progression of ideas that students should develop throughout their K–12 education. Instruction supports students in building more sophisticated ideas by connecting to preceding ideas. The DCI LS1.C: Organization for Matter and Energy Flow in Organisms explores the question "How do organisms obtain and use the matter and energy they need to live and grow?" The DCI LS2.B: Cycles of Matter and Energy Transfer in Ecosystems explores the question "How do matter and energy move through an ecosystem?" The DCI PS3.D: Energy in Chemical Processes and Everyday Life explores the following two questions: "How do food and fuel provide energy?" and "If energy is conserved, why do people say it is produced or used?" These DCIs need to be developed throughout the education of the learner.

The following sections describe some of the major ideas that students will need to develop in grades K–12 and discuss their importance in building the Five Big Ideas of energy over time.

Grades K–2

By the end of second grade, all students should develop usable knowledge that all animals need food to live and grow. Animals obtain their food from plants or from other animals. Plants need water and light to live and grow.

These ideas are necessary precursors to the understanding that all organisms need and use energy for various life functions and growth. These initial ideas help build students' understanding that energy can be converted from one form to another (Big Idea 2) and that energy can be transferred between objects (Big Idea 3).

Grades 3–5

The idea that energy released by burning fuel or digesting food was originally energy from the Sun is also introduced in this grade band, as is the idea that energy can be transferred from fuels and sunlight. Further, students learn the idea that plants capture energy from the Sun to convert raw materials from air and water into food.

By the end of the fifth grade, students further develop the idea that animals and plants alike need to take in air and water. They also explore the idea that animals must take in and digest food. Food provides animals with material for growth and repair and with the energy input required to stay warm and move. Plants, on the other hand, need light, water, and minerals to make the food that they use for growth and repair. Students learn that plants capture energy from sunlight; however, the details of photosynthesis are not yet introduced.

At this grade band, students begin to use the term *energy*. Students should connect the idea of energy to phenomena such as moving, warming, and growing, and they can begin to recognize that energy can be transferred between objects via sound, light, heat, electricity, and collisions. Also at this grade band, students are being introduced to ideas that that all energy is fundamentally the same and that it can be manifested in different forms (Big Idea 1). They also build deeper ideas that energy can be converted from one form to another (Big Idea 2) and transferred from one place to another (Big Idea 3), and they learn that an energy input (i.e., transfer) is necessary for the growth of organisms.

Grades 6–8

By the end of eighth grade, students develop an understanding that plants, algae, and many microorganisms undergo the process of photosynthesis by capturing energy transferred via light to make sugars (food) and oxygen from carbon dioxide and water. Students deepen their understanding that energy can be converted from one form to another (Big Idea 2) and that energy can be transferred between systems and objects (Big Idea 3). Understanding of the importance of sugars for immediate use or storage of sugars for growth or later use is also deepened.

Students in this grade band learn that animals obtain food from eating plants or other animals. Through a series of chemical reactions that occur in the cells of an organism, food is broken down and rearranged to form new molecules to support growth or release energy. This grade band lays the foundation for understanding photosynthesis and respiration as chemical processes that depend on the cycling of carbon in the environment. It is at this grade band that students explore the idea that plants capture and transfer energy from the Sun through a chemical process in which carbon dioxide and water combine to form complex food molecules (sugars) and release oxygen molecules.

Grades 9–12

By the end of 12th grade, student understanding should deepen to include the idea that photosynthesis and cellular respiration provide energy for all of life processes.[7] The ideas move beyond middle school in that students now come to understand that photosynthesis is a complex process and that a series of complex reactions are responsible for the uptake and release of energy in the body. Students also learn that the carbon-based molecules formed during photosynthesis contain carbon, oxygen, and hydrogen and that these carbon-based molecules are responsible for the formation of other carbon-based molecules (e.g., amino acids, DNA) in our bodies necessary for growth and repair. Moreover, students develop the idea that a continuous flow of energy is necessary to sustain life on Earth and that this energy is almost exclusively from the Sun. Students' understanding that energy is conserved—that energy is never created or destroyed, but is only converted or transferred (Big Idea 4)—is further developed by tracing energy flows through organisms and ecosystems.

By the end of 12th grade, students should be able to relate the energy transfers in photosynthesis to those in constructed devices such as solar cells, which capture the Sun's energy and produce electrical energy. By tracing energy flows from the Sun through both natural and designed systems, students further develop the idea that energy can be converted from one form to another and that energy can be transferred between systems and objects. When tracing energy flows, students should consider the efficiency of related energy transformation and transfer processes and use the idea of energy dissipation to account for differences in efficiency (Big Idea 5).

Teaching to Promote an Understanding of Photosynthesis and Respiration

The *Framework* and the *NGSS* stress the importance of students using evidence as the basis for the development of their ideas. In the following sections, we describe several tasks that students can perform that will provide them with an evidence base for building more sophisticated understanding of photosynthesis and respiration, rather than just factual knowledge about the process. Many of the activities in the following sections are modified from or inspired by the IQWST (Investigating and Questioning our World through Science

---

7 Scientists were recently shocked to find that some organisms, referred to as *extremophiles*, live in unexpected places such as the deep ocean floor (where there is no penetration of sunlight) and do not rely on photosynthesis. Instead, they use chemicals from the Earth's interior that are released in ocean vents to carry out chemosynthesis to survive.

and Technology) eighth-grade chemistry unit "How Does Food Provide My Body With Energy?" (Krajcik and Drago 2013).

Grades K–2

At the K–2 level, students should observe that living organisms need to eat to survive. By going to the zoo or a farm, reading about animals, and observing animals around the school (e.g., fish, lizards, beetles) and at home (e.g., dogs, cats, birds), students can identify a pattern—all animals eat. Do they all eat the same things? What do cows and goats eat? What do cats and wolves eat? What do people eat? Noticing that all animals must eat becomes an early precursor to the idea of the flow of energy through ecosystems.

In the early grades, students should collect evidence that plants need water and sunlight. Students can try to grow wheatgrass with and without water. In doing so, students will see a pattern that grass won't grow without water. Students can also observe that sunlight is needed to grow a plant by trying to grow wheatgrass in both light and dark areas and describing the difference in the two conditions after 1–3 weeks. At this grade band, students can also experience that the sunlight can heat various surfaces (Big Idea 3).

Grades 3–5

As students advance into the upper elementary grades, they can continue to experience energy-related phenomena in wheatgrass by observing that it can be dried and burned to heat water or turn a pinwheel that is held above it. By using models like flowcharts, students can trace the energy released when the wheatgrass burns to energy that is transferred from the Sun via light. Students can similarly track and model energy transfers (Big Idea 3) and conversions (Big Idea 2) in light-powered devices such as solar-powered water heaters.

By focusing on straightforward energy transfers and conversions in natural and designed systems, upper elementary students become well positioned to develop a more sophisticated evidence base for energy flows in living systems as they begin the middle grades.

Grades 6–8

At the middle school level, students are ready to transition from focusing on the needs of plants and animals to begin exploring specific evidence that plants use energy from light to produce food (sugars) from carbon dioxide and water, and that this food can react with oxygen to provide energy for plants and animals to live. In this section, we present tasks that students engage in to provide evidence that plants use light to store chemical energy (Big Idea 2), that photosynthesis produces food and oxygen, and that photosynthesis requires an input of carbon dioxide from the atmosphere (Big Idea 3). It is critical to provide middle school students with experiences that allow them to track evidence for photosynthesis.

What Evidence Do We Have That Plants Use Light to Store Chemical Energy?

At the middle school level, students can gather evidence that plant matter stores energy (in the form of chemical energy), which was transferred to the plant via light and can be released (i.e., transformed into thermal energy and light, both of which are transferred to the surroundings) through burning (i.e., reacting with oxygen). To do this, students can let wheatgrass plants grow in two different conditions: exposed to sunlight and not exposed to sunlight (e.g., in a classroom cabinet). Let the plants in each condition grow for 1–3 weeks, then dry the products. Students will obtain more drastic results the longer the plants are allowed to grow. Students should compare their observations of burning the dry wheatgrass grown in sunlight with those of burning dry wheatgrass grown in a dark area.

> **Safety Note**
> Use caution in working with an active flame. It can burn skin and clothing, so keep hands and clothing away from the flame.

The wheatgrass plant that is kept in the dark, while able to grow somewhat, does not have access to energy transferred from sunlight. This plant can use food molecules stored in the wheatgrass seed to provide raw materials necessary for building structures and carrying out cellular respiration. However, with no energy input from sunlight, the plant's growth and development are limited by the number of food molecules stored in its seed. By the end of the experiment, the plant will have exhausted its store of food molecules and no longer be able to continue building structures necessary for growth, which is why students will observe that the wheatgrass grown in the dark has weaker stalks. Without access to sunlight, the plant can't make more food molecules through the process of photosynthesis.

These questions can help students make sense of the wheatgrass experiment:

- *What energy conversions were involved in burning the wheatgrass?* Chemical energy was converted into light and thermal energy, as well as kinetic and sound energy (Big Idea 2).

- *What was the original source for this chemical energy?* Students will have evidence from their experiment that wheatgrass needs light to grow. This is an energy transfer via light (Big Idea 3). The wheatgrass converted the light energy into chemical energy, which was then converted to light, thermal, kinetic, and sound energy during burning.

- *Why did the wheatgrass grown in the sunlight burn more than the wheatgrass grown in the dark?* The wheatgrass in the sunlight had access to an energy input via light (Big Idea 3), which allowed it to photosynthesize and form more structures with a higher-energy molecular configuration. When burned, chemical energy in the plant-air system was converted and released as thermal and light energy (Big Idea 2). Because the plant grown in the sunlight had formed more structures than the plant grown in the dark, it could release more energy when it was burned compared to the plant grown in the dark.

What Evidence Do We Have That Plants Produce Carbohydrates as a Result of Photosynthesis?

Although some of the sugars that a plant produces during photosynthesis are used for immediate energy needs, most of these sugars are stored in the plant for future use (either for growth or energy needs). Sugar molecules such as glucose ($C_6H_{12}O_6$) can join together to form more complex molecules. Two sugar molecules can join together to form a molecule called a *disaccharide*.[8] Figure 5.5 shows the formation of a disaccharide, a process that also releases water. Additions of other sugar molecules form more complex molecules called complex carbohydrates. While a carbohydrate is any molecule containing only carbon, hydrogen, and oxygen, the term *complex carbohydrate* refers to carbohydrates made of a long chain of sugar molecules (see Figure 5.6).

Figure 5.5. Formation of a disaccharide—a simple carbohydrate

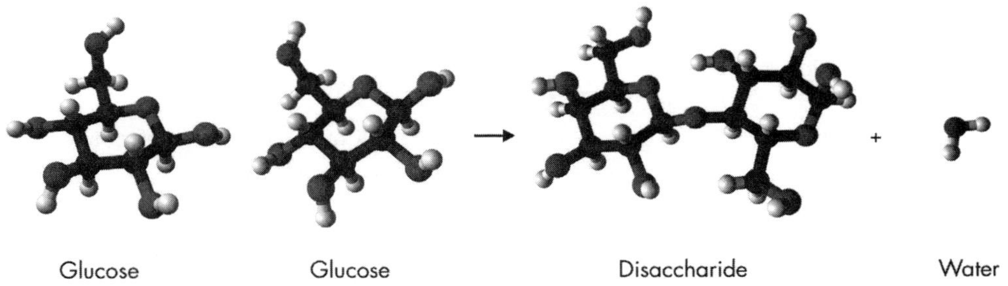

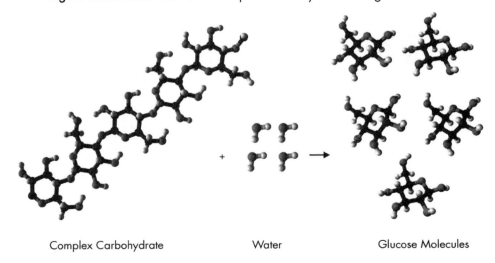

Glucose   Glucose   Disaccharide   Water

Source: Adapted from Krajcik and Drago 2013. Used with permission.

Figure 5.6. Breakdown of a complex carbohydrate into glucose molecules

Complex Carbohydrate   Water   Glucose Molecules

Source: Adapted from Krajcik and Drago 2013. Used with permission.

---

8   The word *disaccharide* means two (di-) sugars (-saccharide). If glucose exists on its own, we can call it a *monosaccharide* (one sugar).

Glucose molecules can serve as carbohydrate subunits and can link together and release water to form starch, a long molecule made up of many glucose subunits. This is similar to what happens when a disaccharide is formed (Figure 5.5), except that many glucose units are added to form a long chain. Plants store these starch molecules to be later broken down to release energy for growth and repair. Figure 5.6 shows the breakdown of starch, a more complex carbohydrate, into glucose (a simple carbohydrate).

Students can conduct a simple experiment to gather evidence that plants use light to produce food and store this food as starch. To set up the experiment:

- Have students grow a geranium plant or provide them with one. (Geranium plants work well for this activity because their leaves stain nicely.)
- Cover some of the geranium leaves completely with aluminum foil to prevent any light from entering.
- Have students grow the geranium plant with some of the leaves covered in aluminum foil for four or five days.

After letting the plant grow for a few days, you can test to see if the leaves contain starch by using iodine. To show students how iodine functions as a detector of starch (food), take a cracker and place a drop of iodine on it. Help students notice how iodine turns a dark purple color in the presence of starch and is yellow in the absence of starch.

Drop some iodine on the leaves that were growing in the light. Next, uncover the leaves wrapped with the aluminum foil and drop some iodine on them. You will see that the leaves grown in the light stain a dark purple color, whereas the leaves wrapped in the aluminum foil do not turn colors and still look green. The leaves that were kept covered with aluminum foil will have less starch than uncovered leaves, because the leaves covered with foil do not have access to the light necessary to undergo photosynthesis and the subsequent synthesis of starch. Leaves in the light carry out photosynthesis and produce glucose that they store as starch, whereas leaves in the absence of light cannot make glucose and will have no extra glucose to store as starch. Leaves in the dark can transform the chemical energy in the starch already stored in the plant for some growth and life function, but only for a limited amount of time.

Safety Notes
1. Review the safety data sheet for iodine with students.
2. Use sanitized indirectly vented chemical-splash goggles, gloves, and aprons during pre-lab setup, during the activity, and during post-lab cleanup.
3. Caution students not to get iodine on their skin or clothing—it will stain.
4. Wash hands with soap and water after completing the lab.

What Evidence Do We Have That Plants Produce Oxygen?

Elodea, a plant that grows in water, carries out photosynthesis like other plants. By closely observing elodea (perhaps with a magnifying lens), students should notice bubbles coming from elodea in water. Ask students what they think these bubbles might be (by this age, students have likely heard that plants give off oxygen). To have students test whether these are, indeed, oxygen bubbles, have students place some elodea in a 250 ml flask and submerge the elodea in water. Now, place a dissolved oxygen probe[9] in the water in which the elodea plant is growing, seal the flask using some aluminum foil, and record the oxygen concentration for 30 minutes. Notice what happens to the oxygen reading. Because it is a closed system, the oxygen level will increase. This provides evidence that the elodea plant is undergoing photosynthesis because oxygen is released. As a control, students should place a dissolved oxygen probe in a test tube filled just with water, and they should notice that the oxygen reading does not change.

What Evidence Do We Have That Plants Use Carbon Dioxide for Photosynthesis?

Just as it is important for middle school students to gather evidence that plants produce food and oxygen, students should also gather evidence that plants must take in carbon dioxide during photosynthesis. To gather evidence for the presence of carbon dioxide, students can use a substance called bromothymol blue (BTB), which is a qualitative indicator for carbon dioxide dissolved in water. You can use BTB to determine whether carbon dioxide is dissolved in water and to get a qualitative sense of just how much is present. A BTB-water solution is blue when little or no carbon dioxide is present; if you add carbon dioxide, the solution becomes greenish and then yellow.

To establish that the solution changes color in the presence of carbon dioxide, you can add carbon dioxide to a BTB solution by simply blowing through a straw into a test tube or a beaker containing BTB-water solution. In the process, the solution will turn from blue to green to yellow as more carbon dioxide is added to the solution. Alternatively, you can add carbon dioxide by using a carbon dioxide cartridge.

To set up an experiment to test whether plants take in carbon dioxide, blow into a beaker containing water and BTB-water solution until it turns a deep yellow color. Then, separate the fluid into two test tubes. To conduct the

> **Safety Notes**
> 1. Use sanitized indirectly vented chemical-splash goggles, gloves, and aprons during pre-lab setup, the activity, and post-lab cleanup.
> 2. Review Safety Data Sheet for Bromothymol Blue (BTB) with students.
> 3. Caution students not to suck in the liquid when using the straw.
> 4. Use caution in working with an active flame. It can burn skin and clothing, so keep hands and clothing away from the flame.
> 5. Make sure the site is free of flammable liquids.
> 6. The bell jar can be dangerous to handle when it is heated. Wait for it to cool down before handling.
> 5. Wash hands with soap and water after completing the lab.

_____

9 These probes can be purchased from science supply companies (e.g., Vernier).

experiment, take a string of elodea and add it to one of the test tubes; leave the other test tube empty (BTB-water solution, but no elodea) so that it serves as a control. Put both test tubes in light; you will see that the liquid in the test tube with the elodea will gradually change from a yellow to a bluish-green color. This color change indicates the removal of carbon dioxide from the BTB-water solution. In your classroom, make sure you have students predict what they think will happen and also have them construct an evidence-based explanation.

You can extend this activity by challenging students to come up with a procedure for testing whether light affects the production of carbon dioxide. Here is one possible procedure:

1. In addition to the two test tubes described above (both filled with a BTB-water solution, one with elodea and the other without), include a third test tube containing a BTB-water solution with the same initial color as the other two, and add elodea to it.

2. Wrap the third test tube entirely with aluminum foil so that no light can enter. Students can poke a small hole in the aluminum foil if they feel it is important, to convince themselves that the third test tube has access to outside air like the other two do.[10]

3. Take all three test tubes and place them in the light.

4. After enough time passes for a color difference to arise between the uncovered test tubes, have students observe each test tube and figure out what the result of each condition means. They should conclude that this activity has provided additional evidence that photosynthesis requires both light and carbon dioxide.

Based on the evidence students gather in the tasks described in this section, middle school students should be able to build a model of the photosynthesis process that illustrates the role of light, carbon dioxide, water, and oxygen as plants produce food. They should also be able to provide an evidence-based explanation of the photosynthesis process. Obtaining evidence is critical for students to build an evidence-based understanding of photosynthesis.

Grades 9–12

Students in high school can carry out more in-depth experiments to collect further evidence to support a more sophisticated model of photosynthesis and respiration. In this section, we present activities that help students gather evidence that plant matter releases

---

10 This hole is not really necessary, because the aluminum foil will not be airtight. As an alternative to wrapping the third test tube in aluminum foil, students could simply place the third test tube—uncovered—in a dark cabinet.

carbon dioxide and water as it burns and a series of tasks that establish that plants use oxygen as they respire.

What Evidence Do We Have That Plant Matter Produces Carbon Dioxide and Water When It Burns?

Students can burn vegetable oil and collect and record the mass of oil, the amount of oxygen, the amount of carbon dioxide, and the amount of water vapor. Vegetable oils such as canola oil, corn oil, and palm oil are complex carbon-containing compounds that plants produce. To measure the amounts of oxygen, carbon dioxide, and water vapor, students can burn the oil in a sealed bell jar and use commercial oxygen, carbon dioxide, and humidity probes[11] to collect data. Once the probes are placed in the bell jar and the oil is lit, the bell jar needs to be sealed with clay. Using real-time software, students can observe that as the oil burns, the amount of oxygen in the bell jar decreases while the amount of carbon dioxide and water vapor (as measured by the humidity probe) increase. Once the flame goes out and the system cools (the flame will go out on its own as the oxygen in the bell jar is depleted), the students can weigh the remaining oil to see if the amount of oil has decreased. Students should be challenged to describe what their data indicate about the burning reaction and how energy is transformed and transferred when plant material is burned.

What Evidence Do We Have That Plants Use Oxygen and Produce Carbon Dioxide During Cellular Respiration?

To investigate this question, students can collect and analyze data using elodea plants. Place an elodea plant and water in a test tube that contains an oxygen probe. Place the test tube in a totally dark environment (e.g., a cardboard box or classroom cabinet) and collect data overnight. The amount of dissolved oxygen will go down because the plant is using its stored food (starch) and oxygen dissolved in the water to make carbon dioxide and water.[12]

As a part of this investigation, students could also monitor the amount of carbon dioxide in the test tube water using BTB. Because the elodea plant is respiring (and thereby producing carbon dioxide), the BTB in the water should also turn the solution more to the greenish-yellow color. Overall, students should gather specific evidence that plants take in oxygen and produce carbon dioxide when they are in a dark environment, just like animals do. This is evidence that plants also undergo cellular respiration.

High school students can also investigate the growth of geranium plants in a sealed bell jar. To set up the experiment, place oxygen, carbon dioxide, and humidity probes in a sealed bell jar. Over time, the amount of carbon dioxide in the jar will decrease and the amount

---

11 These probes can be purchased from science supply companies (e.g., Fourier Education, Pasco, and Vernier).

12 Flowering plants, such as geraniums, generally take water from the roots into their vascular system. Elodea is a flowering plant, too, and has all plant parts, including roots; however, it can also take in water through its leaves.

of oxygen and humidity will increase if the plant is in the light. This happens because the rate of cellular respiration occurs at a much slower rate than the rate of photosynthesis in a plant. Photosynthesis needs to occur at a much faster rate, because material is needed for both growth and energy needs and photosynthesis can occur only in the sunlight.

If students collect data over several days, they will see differences throughout a 24-hour period (i.e., light and dark). Students should be challenged to consider what their data indicate about energy flow in plants in the light and in the dark. To help students make sense of the data, they could draw models of what is going on when a plant is in the light and when it is in the dark. The models should show that in the presence of light, plants use carbon dioxide and water to produce glucose and oxygen (but they also use some stored glucose and oxygen for respiration). The models should not show the production of glucose and oxygen in the dark but, instead, should show the use of glucose and oxygen to form carbon dioxide and water. Table 5.1 summarizes what happens in plants in light and dark conditions and the observations students can make about differences in the two conditions.

Table 5.1.
DIFFERENCES IN PLANT PROCESSES UNDER LIGHT AND DARK CONDITIONS

Process	Plant in light condition	Plant in dark condition
Cellular respiration	Releases carbon dioxide and water but at a smaller rate than they are used up in photosynthesis	Releases carbon dioxide and water
Photosynthesis	Produces glucose and oxygen	Does not produce glucose or oxyen

Students can use a BioChamber (available from Vernier; see Figure 5.7 for an example of a BioChamber setup) with a carbon dioxide gas probe and an oxygen gas probe and software to measure and graph the carbon dioxide and oxygen released during photosynthesis of fresh, turgid spinach leaves. By wrapping the BioChamber in aluminum foil, students can compare the amount of photosynthesis that occurs in the light and in the dark. Students can also explore if the amount of light shining on the leaves has an impact on the amount of carbon dioxide used and oxygen released. Students should be able to use their models to make predictions about what will happen in these conditions. Moreover, the data should reinforce the models that they built. In light, plants convert solar energy into stored chemical energy (i.e., the glucose and oxygen system; Big Idea 2). In the dark, plants use starch to undergo respiration (Big Idea 2). In both cases, energy is not created or destroyed but is transformed. However, in both cases much of the energy tends not to be used (Big Idea 5).

Figure 5.7. BioChamber setup from Vernier with CO_2 gas and O_2 gas probes

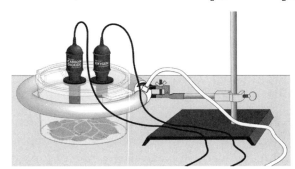

Source: © Vernier Software & Technology. Used with permission.

How Do We Know That Living Organisms Undergo Cellular Respiration?

To explore cellular respiration in non-photosynthesizing organisms, students can explore and describe the relationship between carbon dioxide and oxygen levels in fresh and sterile compost. Fresh compost has numerous microorganisms in it, and students can use carbon dioxide and oxygen probes to measure changes in these gases when the compost is placed in a closed environment. In a container of fresh compost, students will see that the oxygen level increases and the carbon dioxide level decreases over time. In an equal amount of sterile compost, the carbon dioxide and oxygen levels will remain the same (compost contains plant material, but none of it is living so it will neither photosynthesize nor respire). The differences between these two conditions can be explained by the presence of microorganisms in the fresh compost and the absence of microorganisms in the sterile environment. The microorganisms in the fresh compost digest and use many carbon-containing compounds for growth, repair, and other life functions. To release stored chemical energy in the plant material, the microorganisms must take in oxygen to perform cellular respiration reactions; these microorganisms must also release carbon dioxide in the process.

> **Safety Notes**
> 1. Know the source of the compost. Make sure there is no meat or other sources for potentially hazardous microbes to grow.
> 2. Wash hands with soap and water after working with compost.

Once data are collected, students should produce models to explain why the oxygen and carbon dioxide levels behaved differently in the two conditions and relate these changes to the energy inputs (Big Idea 3) and conversions (Big Idea 2) that organisms need to live and reproduce.

In another task, students can use carbon dioxide gas probes to monitor the carbon dioxide produced by peas as they undergo cellular respiration.[13] Students can test germinating and nongerminating peas and look for differences between them. Additionally, students can monitor the products of cellular respiration in germinating peas at two different temperatures.

---

13 See *www.vernier.com/files/sample_labs/BWV-11B-COMP-cell_respiration_CO2.pdf* for complete directions on how to carry out this experiment.

Because germinating seeds are growing but not yet photosynthesizing, the carbon dioxide level will decrease over time. Students can also use oxygen gas probes and observe the oxygen level increase over time. These experiments will provide further evidence regarding what is occurring during respiration.

Summary

Plants make life on Earth possible. Even though the Sun transfers an incredible amount of energy to Earth via light (Big Idea 3), this energy would simply go to heating the planet and be reflected and re-radiated into space if it weren't for photosynthesis. This process makes it possible for plants to transform light energy from the Sun into chemical energy as they build glucose with a higher potential energy arrangement of atoms (Big Idea 2).

Plants get the raw materials for building glucose molecules from carbon dioxide and water in the environment, and these glucose molecules (and other complex carbon-based molecules made by plants) can later react with oxygen in a process that releases the potential energy in the glucose-oxygen system as thermal energy associated with faster-moving particles. Whether we use plant material for food (through cellular respiration) or fuel (through burning), the energy released (i.e., transferred from the food-oxygen or fuel-oxygen system) can be traced directly back to light from the Sun (Big Idea 1).

Just a small fraction of the energy transferred from the Sun to Earth via light is actually captured by plants, and an even smaller fraction ultimately becomes available to the animals that eat the plants or the devices that use fossil fuels to operate. Just like in the cheese puff lab, the missing energy is not gone (Big Idea 4); instead, virtually all of it is transferred to the environment via heat (Big Idea 5). As we have seen in this chapter, both matter and energy change during photosynthesis, cellular respiration, and burning. The matter and energy changes that occur as energy flows through ecosystems and social systems have tremendous social consequences (see Chapter 7).

Photosynthesis makes all life on the planet, including ours, possible (except for the very small number of extremophiles). Yet, few students see the wonder and beauty in photosynthesis and cellular respiration and appreciate how all life on Earth is dependent on plants capturing and transforming energy from the Sun. Students often miss this wonder and beauty because they never get to engage in figuring out what is occurring and thinking deeply about the various matter and energy changes that occur during photosynthesis and cellular respiration (nor do they have the evidence to support their ideas).

Rather than just being presented with the facts of photosynthesis and respiration, students should wrestle with questions such as these:

- Where do you get your energy to do stuff?
- Why does eating food provide you with energy?

- Where did the energy in your food come from?
- What evidence do you have that photosynthesis uses carbon dioxide and produces glucose?
- What evidence do you have that cellular respiration uses glucose and produces carbon dioxide?

Questions like these are simple enough for students to begin exploring at young ages; over time, students can begin to develop more sophisticated answers to these questions as they gather an increasingly rich evidence base for the role of photosynthesis and respiration in living systems. At each grade band, students incorporate this evidence into increasingly sophisticated models that can be used to address these fundamental questions about where life on Earth gets the energy to exist.

References

Carlowicz, M. 2009. NASA satellite detects red glow to map global ocean plant health. *www.nasa. gov/topics/earth/features/modis_fluorescence.html.*

ConsumerReports.org. 2014. Wattage calculator. *www.consumerreports.org/cro/resources/images/video/ wattage_calculator/wattage_calclulator.html.*

Fortus, D., D. Grueber, J. C. Nordine, J. Rozelle, C. Schwarz, D. Vedder-Weiss, and A. Weizman. 2013. Can I believe my eyes? In *Investigating and Questioning Our World Through Science and Technology,* eds. J. S. Krajcik, B. J. Reiser, D. Fortus, and L. Sutherland. Greenwich, CT: Activate Learning.

Krajcik, J., and K. Drago. 2013. How does food provide my body with energy? In *Investigating and questioning our world through science and technology,* ed. J. Krajcik, B. J Reiser, L. Sutherland, and D. Fortus. Greenwich, CT: Activate Learning.

Lindsey, R. 2009. Climate and Earth's energy budget. *http://earthobservatory.nasa.gov/Features/ EnergyBalance/page1.php.*

National Research Council (NRC). 2012. *A framework for K–12 science education: Practices, crosscutting concepts, and core ideas.* Washington, DC: National Academies Press.

National Academy of Sciences. 2015. What you need to know about energy. Our energy sources: Fossil fuels. National Academy of Sciences. *http://needtoknow.nas.edu/energy/energy-sources/fossil-fuels.*

NGSS Lead States. 2013. *Next Generation Science Standards: For states, by states.* Washington, DC: National Academies Press. *www.nextgenscience.org/next-generation-science-standards.*

CHAPTER 6
CONSERVATION OF ENERGY

DAVID FORTUS

Me, My Wallet, and My Money: A Short Story

I have $457.25 in my wallet right now. If my wallet remains in my pocket, I have every reason to believe that there will still be $457.25 in it when I get home in the evening and when I get up and dress tomorrow morning. Indeed, until I take out some money from my wallet or add some money to it, the total amount of money in it will remain $457.25.

I'm hungry, so I buy myself a donut and a cup of coffee for $2.75 and pay for them with the money in my wallet. There's now $454.50 remaining in my wallet. However, the total value of the donut, the cup of coffee, and the money in my wallet is still $457.25. Thus, one could say that while some of my money has changed form (been transformed) from paper bills and coins into a donut and a cup of coffee, the total value remains unchanged.[1]

I meet a friend who owes me $100.00. He gives me a check for $100.00, which I fold and stick in my wallet. I now have a total of $554.50 in my wallet, some of it in cash and some as an undeposited check. One could say that my friend transferred some money to me, so we could say that some money was transferred into the system consisting of me (and my wallet).

I'm flying later in the day to Germany and want to have some local currency (euros) on me when I arrive, so I go to the bank and convert $400.00 into euros. I know that $400.00 is equivalent to 300.00 euros, but the clerk gives me only 285.00 euros because there's a 5% transaction fee. So, I now have on me $154.50 and 285.00 euros. Again, some of my money has been transformed (i.e., from dollars to euros), but this time some money (i.e., the transaction fee) was lost in the transformation process.[2]

The amount of money I have in my wallet limits what I can do. If I want to spend two nights at a hotel in Germany, go out to nice restaurants, rent a car, and bring back a gift for my spouse, I may not have enough money. So, I have three choices: I can do only some of these things, I can go to cheaper hotels and restaurants, or I can find a way to get some more money.

---

1 For the sake of simplicity, we are ignoring important economic factors such as depreciation or the changing value of currency over time.

2 Of course, this "lost" money isn't gone—it was transferred to a much larger system of the bank when they exchanged my money.

The Conservation of Money

Assuming we ignore economic issues such as inflation and shifting currency exchange rates, the total amount of money I have in my wallet is constant unless I spend some or receive some. This is an example of what could be called the conservation of money: Within a given system the amount of money is constant unless some money leaves the system or enters it from outside. In the example given here, the system being considered is my wallet. The money in my wallet can have several forms: For example, it can be in bills, coins, dollars or some other currency, or checks. But unless something enters or leaves my wallet, the total amount of money in it remains unchanged.

I can choose to define a different system other than just my wallet. For example, I can choose a system that is composed of my friend and me (and my wallet). Assume that my friend and I share a joint bank account. In this case, when my friend gave me a check for $100.00, no money actually left or entered the system. The check was relocated in the system, but it did not cross the boundaries of the system. Thus, the total amount of money in the system consisting of my friend and me (and my wallet) remained constant when he gave me the check—or, to put it another way, it was conserved. Note that the total amount of money in the system consisting of *only* my wallet was not conserved during this transaction because some money crossed the system's boundaries—it entered my wallet.[3]

Likewise, when I bought the donut and coffee, the total value was conserved in a system that consists of my wallet and my physical possessions. In a system consisting of my wallet, my physical possessions, my friend, and the bank, the total value of money is conserved throughout the story.

Money is useful to me only if it can change its form. Thus, having euros in my wallet is practically useless to me in the United States because I can't change its form (e.g., change it into donuts and coffee)—I can't use euros to buy anything in just about any U.S. store. To make money useful, sometimes it needs to be transformed a number of times, and in this process sometimes some of the money is lost; it is transferred away from me to something else. Thus, for example, I can't use a check unless it is first transformed into cash at a bank, and there is often a fee for this transformation, just as some money was transferred away from me to the bank when there was a transaction fee for changing dollars to euros.

To summarize, although money can be transformed (e.g., changed to different currencies or into physical possessions) and transferred (e.g., moved between my wallet and my friend's wallet), we can say that the value of money is conserved in an isolated system (i.e., one for which money does not cross its boundaries) even though the money may change its form many times in the system or be transferred between subsystems that are contained within our larger system.

---

3 This situation is similar to having both a checking account and a savings account. If you transfer money from one account to the other, the amount of money in each account changes (is not conserved), but the total amount of money that you have remains the same.

What Does All This Have to Do With Energy?

Money was used in the preceding sections as a metaphor for energy. Systems have energy, and when things happen in or to a system, the total amount of energy the system has remains constant (it is conserved—Big Idea 4) unless some energy enters the system from outside or leaves it (has been transferred—Big Idea 3), even though some of the system's energy may have changed its form (been transformed—Big Idea 2). (See Chapters 1 and 2 for a description of the Five Big Ideas.)

The sole reason that energy is of interest to scientists and engineers is that it is conserved. *A Framework for K–12 Science Education* (NRC 2012, pp. 120–121) summarizes the conservation principle in this way:

> *That there is a single quantity called energy is due to the remarkable fact that a system's total energy is conserved. Regardless of the quantities of energy transferred between subsystems and stored in various ways within the system, the total energy of a system changes only by the amount of energy transferred into and out of the system.*

Knowing that energy is conserved allows scientists and engineers to predict which things can and cannot happen in or to systems. The energy conservation principle sets constraints on what can occur in or to systems. If a certain process requires more energy to occur than is available in a system (e.g., I want to buy something that costs more than what I have in my wallet), that process cannot occur unless additional energy enters the system.

Suppose we have a system in which energy can only be manifested in two different forms. If this system is isolated (i.e., no energy crosses the boundary of the system), then any increase in one form of energy must be exactly matched by a decrease in the other. That is, energy is conserved! The following sections explore two examples of this idea in different disciplines.

Two Examples of Energy Conservation

Example 1: Energy Conservation in Earth Science

Consider a plot of land as a system. Light from the Sun reaches this plot of land, transferring energy to it (Big Idea 3). This energy enters the system consisting of the plot of land, increasing the amount of energy in the system. The energy from the sunlight is transformed into thermal energy (Big Ideas 1 and 2) when the light is absorbed by the land. The land gets warmer.

But every object loses energy to its surroundings by emitting infrared radiation—including you (Big Idea 5)![4] The warmer the object is relative to its surroundings, the faster it loses energy through infrared radiation. As the land gets warmer, it loses energy through infrared radiation at a faster rate.

If the rate at which the land gains energy by absorbing sunlight is greater than the rate at which it loses energy through infrared radiation, it will continue to get warmer and warmer, until the rates at which it gains and loses energy become equal. When this happens, the total amount of energy in the land system remains constant (it is conserved—Big Idea 4), because the rate of energy entering the system is exactly equal to the rate of energy leaving the system. So, the temperature of the land increases until the transfer of energy to and the transfer of energy from the land are equal. At this point, the temperature of the land will remain constant.

By knowing the rate at which energy reaches land by sunlight on a given day and the relationship between infrared radiation and temperature, and using the principle of energy conservation, one can estimate the temperature the surface of the Earth will reach in a given place on a given day.

Taking a broader perspective, the entire Earth radiates energy into space via infrared radiation, and the rate of radiation is affected by the composition of the Earth's atmosphere. The principle of energy conservation helps us understand issues related to global warming by relating the rate of energy transferred to the Earth system via sunlight and the rate of energy leaving the Earth system via infrared radiation. Energy conservation is critical to understanding how the Earth's climate has changed, and will change, over time.

Example II: Energy Conservation in Life Science

Consider a meadow where there are grass, rabbits, and foxes. Each of these organisms can be considered a system through which energy enters and leaves. Energy enters the foxes when they eat rabbits; energy leaves the foxes when they are physically active and they transfer thermal energy to the surroundings (Big Idea 3). In addition, foxes can store energy in them as fat[5] or when they grow (Big Idea 1). When a fox does not have enough to eat, it cannot grow or store energy as fat; it can be active only if it uses the energy that has previously been stored as fat (Big Idea 2). If too much time goes by without eating enough, the fox will use up all its stores of fat and other tissues and, thus, will not be able to be active and will die. To remain alive, a regular flow of energy into and out of the fox is required (Big Idea 4).

---

4 Actually, Earth loses energy to the air and the oceans through conduction, but, for the sake of simplicity, we will ignore this.

5 Of course, when we say "store energy as fat," we mean that it requires an energy input to construct a fat molecule, which has a relatively high potential energy arrangement. We can refer to this potential energy as chemical energy. When we "burn fat," there is a reaction involving oxygen that releases this stored potential energy to produce motion, construct new tissues, and carry on other life functions.

The same is true for the rabbits, except instead of eating rabbits like the foxes do, they eat grass. To remain alive, rabbits also need a regular flow of energy into and out of them.

The grass is also a system through which energy flows; it enters the system when sunlight is absorbed by the blades of grass, is stored in the materials from which the grass is made (sugars and more complex molecules), and leaves as thermal energy when the grass respires.

The grass, rabbits, and foxes together can also be considered a system (in this case, it may be called an ecosystem). Energy enters this system as sunlight and leaves it as the thermal energy released by the constituents of the system (Big Idea 5)—the foxes, rabbits, and grass (see Figure 6.1).

Figure 6.1. A sample representation of energy flow in a simple ecosystem composed of just plants, rabbits, and foxes

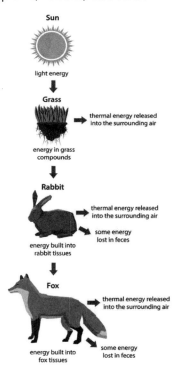

Any activity by one of the constituents of the system releases energy from the system as thermal energy, and the principle of energy conservation requires that, to be able to do so, at least the same amount of energy has to enter the system. Thus, the level of activity that can be maintained in the system is determined by the rate of energy entering the system. This means that by knowing the rate at which energy enters the system (by sunlight), one can estimate the maximum number of foxes and rabbits that can survive in the system and how active they can be (which determines the rate at which energy leaves the system). If there are too many rabbits or foxes that are too active, energy will leave the system at a faster rate than the rate at which it enters the system; this is not sustainable, and if it continues, the organisms in the ecosystem must either become less active or die. One reason why the laws of survival are so unforgiving is that the conservation of energy is an unbendable principle.

What Do Examples I and II Have in Common?

The Earth science and life science examples demonstrate how the conservation of energy plays out in two different systems. They also demonstrate the importance of understanding how the Five Big Ideas of energy (i.e., forms, transformation, transfer, conservation, and dissipation) are involved and intertwined in real-world systems. In considering both systems in our examples, notice how the Five Big Ideas play out:

- **Big Idea 1.** There are always at least two different forms of energy involved in any phenomenon. Example 1 identifies the role of light and thermal energy; example 2 identifies the role of light, thermal, and chemical (potential) energy.

- **Big Idea 2.** Energy is transformed from one form to another (or several others) in such a way that the increase in one form of energy is never larger than the decrease in the other forms of energy, unless there are energy transfers (Big Idea 3) into or out of the system. In example 1, energy transforms from light energy to thermal then back to light (infrared radiation). In example 2, chemical (potential) energy transforms to kinetic energy needed for foxes and rabbits to move.

- **Big Idea 3.** Energy can be transferred into or out of systems, such as when light from the Sun is absorbed by the oceans (example 1) or by the blades of grass (example 2).

- **Big Idea 4.** The constraint described above in Big Idea 2 is due to energy conservation. In example 1, the increase in thermal energy of the Earth is limited by the amount of light transferred to it, which transforms into thermal energy. In example 2, animals are limited in their ability to carry on life functions because of the energy transfers and transformations required.

- **Big Idea 5.** In every non-isolated system, some of the energy is transferred out of the system to its surroundings as thermal energy, where it is spread out or is dissipated.[6] This is why warming is "global" in example 1 and why energy is "lost" to the surroundings in example 2.

---

6 Even in isolated systems, the energy will spread out as much as it can within the system boundaries. But nature doesn't seem to care much about the system boundaries that we draw, which is why we see so many systems that are not isolated.

Building the Idea of Energy Conservation Over Time

The conservation of energy is the most important aspect of the energy concept, but it is also perhaps the hardest to notice since so many of the systems and phenomena are always transferring energy to and from them at different rates and in ways that are not always obvious. Thus, the idea of energy conservation should not be introduced to young children; rather, it should be built over time.

The following sections describe a few of the *Next Generation Science Standards* (*NGSS*; NGSS Lead States 2013) performance expectations (PEs) from different grade bands that help contribute to a developing understanding of energy conservation, together with suggestions for activities that can support the learning of these PEs. There are other energy-related PEs at every grade band, but the ones listed here are some of the PEs that contribute most strongly to the idea of energy conservation.

Grades K–2

At this grade band it is important that students observe phenomena and learn to discuss what they see, hear, feel, or smell. These observations become conceptual anchors on which abstract concepts can be constructed at later ages. Energy conservation is an abstract principle based on abstract concepts, and students at this grade band are simply not ready to deal with these ideas. Energy conservation should not be taught explicitly at this grade band; rather, the goal is to help students develop a repertoire of experiences on which abstract ideas can be built at a later stage.

Make observations to determine the effect of sunlight on Earth's surface (K-PS3-1): Go outside with the students on a bright day and have them feel different surfaces that are in direct sunlight—the sidewalk, asphalt pavement, automobile rooftops, sand—and then have them feel similar surfaces that are shaded. Where do the surfaces feel warmer? Which surfaces felt the hottest? Does the color of the surface make any difference in the sun? In the shade? There is no need to go into explanations; just make sure that the students recognize that the surfaces in direct sunlight feel warmest. This can become a very early precursor to the idea of energy entering or leaving a system through heat transfer.

Use observations to describe patterns of what plants and animals (including humans) need to survive (K-LS1-1): By going to the zoo or a farm, reading about animals, and observing animals around the school (e.g., fish, lizards, and beetles) and at home (e.g., dogs, cats, and birds), students can begin to notice a pattern—all animals eat. Do they all eat the same things? No. What do cows and goats eat? What do cats and wolves eat? What do people eat? This can become an early precursor to the idea of the flow (i.e., transfer) of energy through ecosystems.

Grades 3–5

At this grade band students begin to use the term *energy* as an indicator of activity, such as when something is moving, or as something that is needed for an activity, such as energy needed to warm something or to grow. Students become ready to identify and discuss different forms of energy such as light energy, sound energy, electric energy, and motion energy. Although students begin to discuss ideas such as energy forms, transformations, and transfers, they are still not ready to learn about the conservation of energy. The focus at this grade band is still on connecting energy ideas to familiar phenomena, and these connections deepen the conceptual anchor that students began to build in grades K–2.

Use models to describe that energy in animals' food (used for body repair, growth, motion, and to maintain body warmth) was once energy from the Sun (5-PS3-1): This is a direct continuation of the disciplinary core idea described in the K–2 section above (K-LS1-1) and leads students to take another step toward understanding the idea of energy flow and conservation in ecosystems. If the students know that they eat to get energy and that what they eat got its energy by eating something else or by growing in the sunlight, they can draw flow diagrams to trace the energy they get by eating back to energy from the Sun.

Safety Notes
1. Wear sanitized safety glasses or goggles when working with wires having exposed metal—sharp hazard!
2. Stay clear of water or other liquids when working with live wires—shock hazard!

Apply scientific ideas to design, test, and refine a device that converts energy from one form to another (4-PS3-4): If electrical energy is discussed as the energy that comes from a battery or from a power outlet on the wall, students can investigate that by connecting different types of appliances to batteries or to an outlet from which they can get different kinds of energy—sound energy (e.g., a radio), light energy (e.g., a flashlight), thermal energy (e.g., a radiator), or motion energy (e.g., anything with an electric motor). When asked where these different kinds of energy come from, students will readily acknowledge that they come from the electricity. It is important to emphasize the idea that electricity carries electric energy that can be changed by the appliances (i.e., transformed) into different kinds of energy. This is the beginning of the idea of energy transformation as a precursor to energy conservation—that if one kind of energy increases, another type must decrease.

Make observations to provide evidence that energy can be transferred from place to place by sound, light, heat, and electric currents (4-PS3-2): The appliances in the 4-PS3-4 activity are connected to batteries or to electric outlets with wires. Replace these wires with longer wires or with an extension cord and move the appliances away from the battery or outlet. Show that the appliances still work. How did the energy get to them? The electricity must be moving the energy from the battery or from the outlet to the appliance, so electricity is a way of moving energy from one place to another. A moving object has energy, so when a ball rolls from one place to another, the energy it has moves with it; thus, moving objects must also be another way of moving energy from place to place. This disciplinary core idea is a precursor to the idea of transfer, which is the mechanism by which energy enters or leaves a system.

Ask questions and predict outcomes about the changes in energy that occur when objects collide (4-PS3-3): A rolling red ball strikes a stationary blue ball, causing the blue ball to start rolling and the red ball to slow down. Have students ask questions about this phenomenon. Help them by scaffolding with some questions of your own, such as "Which ball slowed down and which one sped up?" "Does energy increase or decrease when an object speeds up?" "Does energy increase or decrease when an object slows down?" and "Which ball gained energy during the impact?" Because the faster an object moves the more energy it has, the energy of the red ball must have decreased during the collision and the energy of the blue ball must have increased. This simultaneous decrease and increase provides evidence that energy is, in fact, transferred from the red ball to the blue ball during the collision. Later, when students learn how to quantify energy, they can relate the amount by which the red ball's energy decreased to the amount by which the blue ball's energy increased. This analysis will reveal that in addition to transferring energy from one ball to another, this collision must have produced other form(s) of energy as well (e.g., sound and thermal energy).

Grades 6–8

The middle grades make a crucial link between the elementary expectations, which help students connect energy ideas to familiar phenomena, and the high school expectations, which focus on formalizing energy as a quantitative scientific concept. Middle grades instruction is still firmly grounded in phenomena and deepens students' nascent understanding of energy transfer and transformation. With a rich understanding of the different manifestations of energy and how energy changes within a system or moves between systems, students are well positioned to learn about the principle of energy conservation in high school.

Develop a model to describe that when the arrangement of objects interacting at a distance changes, different amounts of potential energy are stored in the system (MS-PS3-2): Gravitational potential energy is the energy stored between objects that apply a gravitational pull toward each other. The gravitational energy of a system of objects increases when the objects get farther away from one another and decreases when they get closer to each other. Thus, when an object is dropped from a tower, the gravitational energy of the system consisting of the object and the Earth gets smaller as the object falls, because the distance between the Earth and the object decreases. Objects can also store elastic energy because of the forces between the particles that make up the object. When the object is stretched, its elastic energy increases because the distances between the particles that make up the object get farther away from one another.[7] Energy is not only a property of individual objects but also

---

7   As long as it is not stretched too far! If the particles get too far apart, they will no longer "bounce back" and the material will be damaged or broken. Even the most stretchy rubber bands eventually snap!

of a system of objects. The idea of potential energy is only relevant for a system of objects—a single object can never possess potential energy.

Construct, use, and present arguments to support the claim that when the kinetic energy of an object changes, energy is transferred to or from the object (MS-PS3-5): When an arrow is shot, its motion energy increases, but, at the same time, the elastic energy of the bow decreases. When an object is thrown up into the air, its motion energy decreases, but the gravitational energy of the object and Earth system increases. If the motion energy of an object could change without any other change occurring, energy would not be conserved. Energy conservation dictates that every change in one kind of energy has to be accompanied by at least one other change (in the opposite direction) in at least one other kind of energy at the same time.

Apply scientific principles to design, construct, and test a device that either minimizes or maximizes thermal energy transfer (MS-PS3-3): When an ice cube is placed in a cup of warm water, the ice cube gets warmer and melts, while the water in the cup get colder. Energy has been transferred from the surrounding water to the ice cube. When a cold hand is placed on someone's warm face, the hand gets warmer and the face colder. Energy has moved from the face to the hand. When a hot oven is turned off, it slowly gets cooler and the surrounding air get warmer. Energy moves from the oven to the surrounding air. These are all examples of thermal energy transfer, also known as heat. Heat is one of the main mechanisms by which energy leaves or enters systems. Give the students a design challenge: Design and test improvements to a mug so that hot chocolate poured into it will cool down slower than normal. Check to see if the same improvements slow down the heating of ice water placed in the same improved mug.

Develop a model to describe the cycling of matter and flow of energy among living and nonliving parts of an ecosystem (MS-LS2-3): This idea is the next step, after 5-PS3-1 (described earlier in this chapter), toward understanding energy conservation in ecosystems. Food webs represent the complexity of energy transfer in real ecosystems and can be used to predict and explain how changes in energy flow (e.g., via changes in sunlight, predation, disease) in one component of the ecosystem can cause a cascade of other changes—all based on the idea that an organism cannot survive unless its energy input meets its energy output. Have students select different ecosystems and draw the flow of energy in the ecosystem, starting from the Sun up to the top predator in the system.

Grades 9–12

In high school, students become ready to learn the quantitative principle of energy conservation. The PE presented here emphasizes that results of the idea of energy conservation can be quantitatively verified.

Create a computational model to calculate the change in the energy of one component in a system when the change in energy of the other component(s) and energy flows in and out of the system are known. (HS-PS3-1): For example, in the toy called Newton's cradle (Figure 6.2), energy is transferred from one ball to another, transformed from kinetic to gravitational and back to kinetic, and from kinetic to elastic and thermal and then back to kinetic. At any given time, the total sum of the kinetic and elastic energy of each ball, plus the gravitational energy of the balls-Earth system, plus the thermal energy that has been transferred to the surrounding air is constant and nonchanging, though the distribution of the total energy among the different types and locations will change.

As the Newton's cradle swings, the balls will slow down and eventually stop moving as the balance of energy gradually shifts to thermal energy in the surroundings—they must stop because energy is conserved. The energy that was originally put into the system when a ball was pulled back is limited, and all of this energy will eventually be transferred back out of the system. Likewise, a fresh battery has a given amount of chemical energy stored in it. When the battery is connected to a lightbulb with wires, the bulb transforms this energy into light and heat, which escape the battery-bulb-wires system. Over time, the amount of energy transferred to the surroundings is exactly equal to the energy transferred from the battery. This is the principle of energy conservation.

Figure 6.2. Newton's cradle

Source: Alsterdrache, *https://commons. wikimedia.org/wiki/File:Kugelstoszpendel.jpg*

Teaching to Promote Understanding of Energy Conservation

This section gives four examples of learning activities—one at each of the four grade bands of K–2, 3–5, 6–8, and 9–12—that can help students develop a deeper cross-disciplinary understanding of energy conservation.

Grades K–2

At this stage of development, it is important to help children see that things can change and that several things can happen at the same time. Students can experience a series of phenomena that highlight this: a cup of water getting colder as the ice in it melts, a yo-yo that spins slower as it rises on a string and then faster as it goes down, a rubber band that gets warmer when it is stretched and cooler when

Safety Note
Wear sanitized safety glasses or goggles during these activities.

it contracts back to its original length, a ball hanging from a rubber band that spins slower as the rubber band get wound more tightly, and so on. Although each of these phenomena can be explained as a result of the transformation of energy from one form to others or the transfer of energy from one system to another, all while maintaining energy conservation, the concept of energy is not to be mentioned! No explicit conclusions need to be reached other than when one thing changes, something else may change at the same time. Do not attempt to make causal connections between the two changes. These phenomena are experiential anchors on which conceptual learning can be built at a later stage.

To be clear, energy as a scientific concept is not to be introduced at this age. The concept only becomes useful after students have noticed the relationship between changes that occur within interacting objects and devices.

As the children play with the objects and observe the phenomena, you can ask questions to lead them to the observations and connections that help set the foundation for learning about energy later on. For example, the following questions would be useful for helping children think about a ball hanging from a twisted rubber band as it spins back and forth:

- How does the ball move?
- Does the ball ever change direction?
- Does the ball sometimes spin faster and sometimes spin slower?
- Does the ball ever stop spinning, even for an instant?
- What happens to the rubber band as the ball spins?
- When the ball changes direction, is the rubber band tightly twisted or loosely twisted?
- Does the rubber band get twisted only in one direction or in both directions?
- If we wait long enough, does the ball stop spinning?
- Does this happen when the rubber band is twisted or untwisted?

These questions focus children's attention on how the ball moves and how these movements are related to the twisting/untwisting of the rubber band. That is, they relate observable changes within systems to one another. Regardless of the phenomenon, prompting children to think about how changes in one thing affect other things will help set the stage for learning productively about energy in the future.

Grades 3–5

The goal at this stage is to help students identify the presence of five different forms of energy in systems (i.e., kinetic, thermal, electric, light, and sound), determine qualitatively when there is more or less of these different types of energy, and recognize that energy can be moved from place to place. Although the term *energy* is introduced as a scientific idea, energy conservation is not a concept that should be discussed at this stage. Students at this age should focus on three of the Big Ideas about energy: energy can exist in different forms (the five forms mentioned above), energy can be changed (transformed) from one form to another, and energy can be transferred from one location or object to another.

Figure 6.3. An electric circuit useful for investigating energy transformations and transfers in upper elementary school

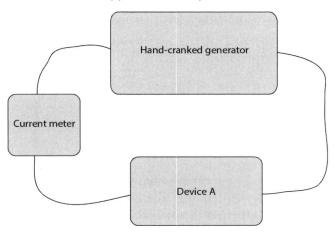

One activity that is particularly useful at this age is to explore electric circuits that are powered with a hand-cranked generator. Figure 6.3 shows a circuit that can be useful for helping students explore identify energy forms, transformations, and transfers in electric circuits.

Device A can be any one of the following items: a lightbulb, a buzzer, a resistor, or another hand-cranked generator (which will turn when the other hand-cranked generator turns, effectively functioning as a motor). Place any one of the items in the circuit as Device A. Have the students crank the generator while observing the lightbulb, hearing the buzzer, feeling the resistor, and seeing the generator or motor spin. The goal is for students to notice the following things and, then, to reach conclusions related to the learning goals for this stage: (a) the items operate only when the generator is being cranked; (b) when the generator isn't being cranked the reading of the current meter is zero; (c) the faster the generator is cranked, the higher the reading on the current meter; (d) the faster the generator is cranked the brighter the light gets, the louder the buzzer gets, the warmer the resistor gets, and the faster the generator or motor spins; and (e) some units are easier to operate than others.

> **Safety Notes**
> 1. Wear sanitized safety glasses or goggles when working with wires having exposed metal—sharp hazard!
> 2. Stay clear of water or other liquids when working with live wires—shock hazard!

Use the following guiding questions to promote classroom discussion about the ideas of energy forms, transformation, and transfer (answers are in parentheses):

- What forms of energy are present when the lightbulb lights? (light energy)
- Where does this energy come from? (from the cranking of the generator)
- How does the energy get from the cranking of the generator to the lightbulb? (electricity)
- What kind of energy is present when the generator is cranked, regardless of what is connected to the generator in the circuit? (motion energy)
- When I crank slowly, is there more motion energy or less? (less)
- When I crank slowly, does the light illuminate brighter or weaker? (weaker) Does this mean that it emits more or less light energy? (less)
- So, less motion energy in the cranking goes together with less energy being emitted by the lightbulb? (correct)
- How do you think the two are connected? (Suggest that maybe the motion energy gets changed into light energy.)

Repeat this sequence with the buzzer, placing it in the circuit as Device A, having students observe what happens, and adapting the guiding questions (and answers) as needed. With the last question, students should reach the same tentative idea, that the kinetic energy in the cranking is being changed (transformed) into sound energy. At this stage, some students will already suggest this possibility. Then, repeat the whole sequence with the resistor. Most students will already accept the hypothesis of energy transformation.

Point out to the students that the energy they are thinking about is in different locations: the motion energy is where their arm and the moving generator are located or where the spinning motor or generator is located, while the light energy, sound energy, and thermal energy (or heat[8]) are located where the lightbulb, buzzer, or resistor are located. If the motion energy is being transformed into other types of energy, it has to get from one location to the others. How does that happen? This is when you direct their attention to the reading of the current meter. Using questions similar to those listed above, help the students conclude that motion energy is being transformed into electrical energy, which allows it to move along the circuit to Device A, where it gets transformed into light, sound, thermal, or motion energy—depending on the device that is attached.

Grades 6–8

At this stage, students need to distinguish between kinetic forms of energy (i.e., motion and thermal) and potential forms of energy (i.e., gravitational, electrical, chemical, and elastic), to recognize that thermal energy always gets transferred from warmer places to colder places and that the increase or decrease of one form of energy is always accompanied by an opposite decrease or increase of at least one different form of energy.

---

8 At this age, it is not necessary for students to distinguish between thermal energy and heat (see discussion of heat in Chapter 4).

Consider the phenomenon described in the "Can I Believe My Eyes?" unit of the IQWST (Investigating and Questioning our World through Science and Technology) curriculum (Fortus et al. 2013), in which electrical energy is transformed into light energy, as described in the next paragraph.

A strong spotlight illuminates two beakers of water (see Figure 6.4); one beaker has just clean tap water, and the other beaker has food coloring added to its water. The colored water warms up faster than the clear water and evaporates faster than the clear water. Electrical energy is transformed into light energy by the spotlight. Some of this light energy reaches the two beakers of water, where some of it is absorbed, reflected, or transmitted by the water. More light is absorbed by the colored water than by the clear water, leading to more light being reflected and transmitted by the clear water than by the colored water. The light that is absorbed by the water transfers energy to the water, where it is transformed into thermal energy. Because more energy is transferred to the colored water than to the clear water, the colored water's thermal energy increases faster than that of the clear water. The increase in the water's thermal energy is accompanied by a rise in the water's temperature. This means that the colored water gets warmer faster than the clear water. As the water gets warmer, its rate of evaporation increases. So, the colored water evaporates faster than the clear water.

Figure 6.4. A spotlight illuminates two beakers of water, and a light sensor measures the amount of light transmitted and reflected through each.

Source: Fortus et al. 2013. Used with permission.

The total light energy reaching any beaker equals the energy of the light being reflected and the energy of the light being transmitted and the energy of the light being absorbed; in mathematical form this is expressed as

Light energy reaching a beaker	=	Light energy reflected from the water in a beaker	+	Light energy transmitted by the water in a beaker	+	Light energy absorbed by the water in a beaker

This is an example of the law of energy conservation: The energy leaving a system and the change of energy in the system must equal the energy reaching the system. It is used regularly by scientists to calculate how the sunlight reaching the Earth heats the Earth's surface.

Grades 9–12

Have students build the following three electrical circuits using three fresh AA batteries and five identical resistors: one battery is connected to one resistor, another battery is connected in series to two resistors, and the third battery is connected in parallel to the remaining two resistors. Each resistor and each battery is placed in a calorimeter and covered by an identical amount of oil. The students should measure the temperature increase in each calorimeter until the batteries are dead. They should then relate the temperature increases to each other and see if the amount of energy originally in each of the batteries (assumed to be the same for each battery) is equal to the amount of thermal energy emitted by the resistors in each configuration.

Safety Notes
1. Wear sanitized safety glasses or goggles when working with wires having exposed metal—sharp hazard!
2. Wipe up any oil spilled on the floor—slip-and-fall hazard!
3. Wash hands with soap and water after completing the lab.

Energy conservation is not readily apparent within the vast majority of macroscopic systems. Thus, rather than try to empirically establish energy conservation in laboratory investigations, students at this grade level should use the law of conservation to produce quantitative models of energy transformation devices or organisms. As discussed in Chapter 3, students can use calorimetry to test whether energy seems to be conserved within a calorimeter and to think about how they can design a better calorimeter that comes closer to the caloric value given on food nutrition labels.

Students can also use the principle of energy conservation to model a combustion engine with the apparatus shown in Figure 6.5. In this investigation, students use a canister of chafing fuel (which rests on top of a balance) to heat the air in an Erlenmeyer flask with a plastic syringe connected to it. A string attached to the syringe plunger runs under a pulley and to a cart that is pulled up an incline in front of a motion detector. Using the heat of combustion for chafing fuel and data from the motion detector, students can calculate the efficiency of their model engine as it pulls the cart up the incline. The efficiency is

Figure 6.5. Apparatus for modeling the efficiency of a (really bad) combustion engine

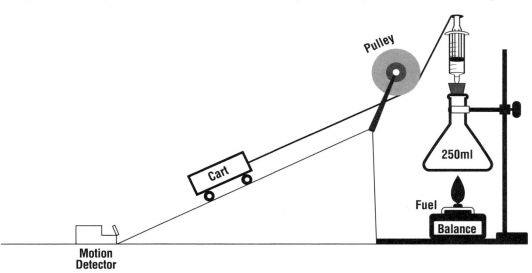

dismal—far less than 1%! Data from this investigation allow students to use the principle of energy conservation to quantify how much of the energy was released by burning the alcohol (which transforms potential energy in the alcohol-oxygen reactant system to kinetic energy of the product molecules), how much was eventually transferred to the cart (which is manifested as increased gravitational and kinetic energy), and how much was "wasted"[9] (i.e., the amount missing from the cart) and compare the efficiency of their device with that of a real internal combustion engine in a car.

As a follow-up to the investigation associated with Figure 6.5, students can build an apparatus to model an electric car (see Figure 6.6, p. 122). In this follow-up investigation, students can use data from current and voltage probes to calculate the power delivered to an electric motor in a certain period of time. Coupled with data from the motion detector, students

---

9 Not all energy missing from the cart is actually wasted—some energy released in the alcohol burning reaction must go to expanding the gas in the Erlenmeyer flask. Even if there were no dissipation at all, this would mean (because of energy conservation) that the energy transferred to the cart must still be less than the energy released as the alcohol burns.

Safety Notes

1. Wear sanitized indirectly vented safety goggles during pre-lab setup, the activity, and post-lab cleanup.

2. Use a long-handle match or butane lighter to light the can. Do not use a lit can to light another can.

3. Never position a can on a table top or on table linen. Never handle or carry a lit can or heating equipment (e.g., a chafing dish) with lit cans inside. Do not leave a burning can unattended.

4. Use caution when working around a heat source, as it can burn skin and clothing. Keep hair, clothing, and all other combustible items away from the flame.

5. Do not handle any equipment when hot—wait for it to cool down. Never use your hands to extinguish a flame; extinguish the flame by blowing it out. Allow wick to cool sufficiently after extinguishing the flame **before** moving the can. Do not touch the wick after the flame has been extinguished. Do not tamper with the wick.

6. Replace the lid/cap before disposal to prevent any fuel leakage.

Figure 6.6. Apparatus for modeling the efficiency of an electric motor

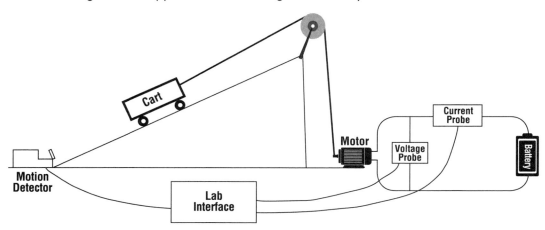

can calculate the efficiency of the electric motor in pulling the car up the ramp. Then students can compare this efficiency with that of their combustion engine and discuss how much energy from the power source was wasted in each scenario (i.e., released from the power source without being manifested in the kinetic or gravitational energy the cart). Students can then expand these efficiency investigations to model energy efficiency for larger systems, such as powering a fleet of automobiles with a certain efficiency. In each scenario, students' models and representations should account for all energy inputs and outputs for the system of interest—that is, they should demonstrate the principle of energy conservation (Big Idea 4). Models should represent how relevant forms of energy (Big Idea 1) change as energy is converted (Big Idea 2) from the source and transferred to the cart (Big Idea 3), while losing a significant fraction of the released energy as thermal energy (Big Idea 5).

Summary

The single most important idea about energy is the fact that it is conserved. However, students are not ready to be introduced to energy conservation as a quantitative principle until high school. The *Framework* and the *NGSS* both recommend that students in grades K–8 focus on energy forms, transformations, and transfers within phenomena drawn from a variety of disciplines to build an understanding of energy conservation over time. In grades 9–12, students should build on their experiences tracking energy transformations and transfers to construct ideas about the quantitative conservation of energy. Of course, energy conservation is almost never apparent through empirical measurements of real systems; thus, the idea of energy dissipation goes hand in hand with that of conservation. Whether students are studying living, chemical, physical, or Earth systems, energy conservation is usually never apparent within a system under study because energy transfers to or from the surroundings are happening all the time. Thus, the concept of systems is fundamentally connected to the idea of energy conservation.

When teaching students about energy conservation, it is critical not to do too much too soon. Even though very young students may be able to repeat the words "energy is never created or destroyed," it is not possible for students to deeply understand the idea of conservation until they have had a rich set of experiences with energy-related phenomena that the energy conservation principle can help them connect together. Throughout grades K–8, students should gradually connect different phenomena to the broad idea of energy and then systematically investigate energy transfers and transformations in a wide range of energy-related contexts from across the scientific disciplines. If students are introduced to the principle of conservation too soon, it may seem like simply a fact to memorize in science class that has little to do with the real world. This is why the *Framework* and the *NGSS* recommend that the idea of energy conservation be built over time as students build an increasingly robust base of evidence that they can use to describe how energy behaves in a range of phenomena. Although conservation is the most central and important idea about energy, it is only after students have a wide range of experiences with the energy-related phenomena that they are ready to understand and use the principle of energy conservation.

References

Fortus, D., D. Grueber, J. C. Nordine, J. Rozelle, C. Schwarz, D. Vedder-Weiss, and A. Weizman. 2013. Can I believe my eyes? In *Investigating and Questioning Our World Through Science and Technology,* eds. J. S. Krajcik, B. J. Reiser, D. Fortus, and L. Sutherland. Greenwich, CT: Activate Learning.

National Research Council (NRC). 2012. *A framework for K–12 science education: Practices, crosscutting concepts, and core ideas.* Washington, DC: National Academies Press.

NGSS Lead States. 2013. *Next Generation Science Standards: For states, by states.* Washington, DC: National Academies Press. *www.nextgenscience.org/next-generation-science-standards.*

CHAPTER 7
ENERGY AND NATURAL RESOURCES

ROBERT CHEN

Why Focus on Natural Resources?

Energy is one of the most important natural resources that the Earth provides for us. In 2010, human civilization consumed 524 quadrillion British thermal units of energy; this usage is predicted to increase by 56% by 2040 (U.S. Energy Information Administration 2013) and could triple by 2100 (Intergovernmental Panel on Climate Change 2000). Energy required for human society comes entirely from Earth's natural resources. These natural resources include nonrenewable fossil fuels such as coal, oil, and natural gas, as well as renewable resources such as wind and solar. How individuals, communities, countries, and humanity as a whole will address this increasing energy demand is one of the greatest challenges of the 21st century. Educating global citizens who are equipped to deal with this grand challenge requires helping students develop fundamental understandings about the energy concept that help them to solve problems, analyze complex systems, and make sense of the world. This crosscutting conceptual knowledge also allows students to make important lifestyle decisions that affect both society and the planet.

As discussed in Chapters 1 and 2, our first Big Idea about energy is "All energy is fundamentally the same, and it can be manifested in different phenomena that are often referred to as different 'forms' or 'types.'" This Big Idea might seem perfectly sensible in the classroom where we can define all energy units used as joules, but other units are seen when dealing with energy bills at home. Therms, kilowatt-hours, gallons of oil, or cords of wood are common units that are not easily converted or understood by consumers, yet everyone understands the bottom line of an energy bill, where these energy units are converted into dollars (see Chapter 6 for a more in-depth analogy between energy and currency). Indeed, while we might not be able to define what energy is or describe all the different units available for measuring it, most of us recognize energy as something that is valuable.

Many students, and indeed many classroom teachers, have trouble identifying the source of energy for operating their coffeepot. "It comes from the wall" and "it comes in from the street" are examples of common responses you might hear from your students.

Although most people in the United States know that cars run on gasoline, a fossil fuel that is dug up from the ground and refined to make it "clean" to burn, many people are less familiar with the array of energy resources used by power plants to generate the electricity that is delivered to our homes via electrical transmission wires.

The U.S. Department of Energy is conducting an Energy Literacy campaign to address this deficiency in public knowledge (U.S. Department of Energy 2014). The department partnered with 13 other federal agencies and more than 20 educational partners in an attempt to increase the fundamental knowledge needed to support critical energy decisions. This effort resulted in the development of a guide entitled *Energy Literacy: Essential Principles and Fundamental Concepts for Energy Education,* which identified seven Essential Principles and a set of Fundamental Concepts to support each principle. The guide, which can be downloaded at *http://energy.gov/eere/education/downloads/get-free-copy-energy-literacy-framework,* states that an energy-literate person

- can trace energy flows through systems,
- knows how much energy he or she uses,
- can assess the credibility of information about energy,
- can communicate about energy in meaningful ways, and
- can make informed decisions about energy based on understanding.

The guide claims that energy literacy is both an understanding of the nature and role of energy in the universe and our everyday lives and the ability to apply this understanding to answer questions and solve problems. An increasingly energy-literate global population is likely to lead to more informed decisions, promote more sustainable energy use, reduce environmental risks, and promote economic development.

The technological systems that humans have designed for energy resource discovery, harvest, and delivery are complex. Understanding these systems requires a deep understanding of the Five Big Ideas of energy as well as the energy concepts as outlined in the *Next Generation Science Standards* (*NGSS;* NGSS Lead States 2013), and such a deep understanding takes many years throughout K–12 education to develop. Although natural energy systems are complex, studying them provides many valuable teaching and learning opportunities because the issues and technologies for harnessing natural energy resources are so common in the news, in students' homes, and in their everyday lives.

The issues involved with identifying and using energy resources present scientific and technological challenges (as well as economic, environmental, and political challenges) that will continue to be a central issue throughout the lives of today's students. These challenges require ideas from Earth science, chemistry, physics, and biology. As such, the study of natural energy resources can provide a rich instructional context that helps students learn to use energy ideas consistently across disciplinary boundaries and to form conceptual bridges between their in-school and out-of-school learning about the energy concept.

Natural Resources From an Energy Perspective

A Framework for K–12 Science Education (NRC 2012) describes the importance of natural resources in this way:

> *All materials, energy, and fuels that humans use are derived from natural sources, and their use affects the environment in multiple ways. Some resources are renewable over time, and others are not. (p. 192)*

To develop a deep understanding of the importance of natural resources, students must develop their knowledge of the Five Big Ideas of energy and how these ideas are related to one another. Energy that is used in the home can take several forms (Big Idea 1), such as electricity, heat, and light, and can be derived from several sources, such as oil, natural gas, coal, wood, wind, and sunlight. These are all fundamentally the same energy, but they can be transformed from one form or type to another by special machines or engines (Big Idea 2). Sometimes this conversion can occur in the home (e.g., a generator can convert gasoline to electricity in case of a blackout), but, many times, the conversion is done at a central plant (e.g., burning coal to make electricity). Energy must be transferred (Big Idea 3) from its source to the end user (e.g., oil well to refinery to gas station to car). As it is transformed or transferred, energy is conserved (Big Idea 4) in that energy is not created or destroyed, but it is dissipated (Big Idea 5) or lost as usable energy for the end user.

The ratio of the amount of energy delivered to the consumer divided by the amount of energy available in a raw natural energy resource is the efficiency of an energy resource harvesting system, and typically this ratio is much less than 1. A very small value for this ratio means that the vast majority of the energy available is dissipated somewhere between harvesting (e.g., mining for coal for use in a coal-fired power plant) and consumer use (e.g., turning on a lamp) of energy. Consideration of the efficiency of energy harvesting and use is critical to conserving our energy resources, but this type of conservation should not be confused with the law of conservation of energy, though many students often confuse the two uses of the term *energy conservation* (Boyes and Stanisstreet 1990). To clarify the two types of conservation, consider using the terms *energy conservation* or *conservation of energy* to refer only to the scientific principle and the terms *energy resource conservation* or *conservation of energy resources* to refer to choices we make that reduce demand for limited of energy resources.

With limited natural energy resources, conservation of energy resources will play a large role in allowing Earth processes to meet the overall societal demand for energy. Driving fewer miles each day, replacing incandescent lightbulbs with LED bulbs, and increasing the fuel efficiency of cars are examples of personal choices and policy decisions that will reduce global demand for the consumption of energy resources. Together with developing new technologies for harvesting energy from nature (e.g., high-efficiency, low-cost solar cells), decisions that reduce demand for energy in our everyday lives will help meet the increasing energy demands of a growing world population.

Learning About Energy Resources Over Time

By focusing intentionally on building the Five Big Ideas of energy over time, teachers can help students develop competencies that are critical for addressing the societal energy challenges of the future and making energy-related decisions in their everyday lives. For example, if a new type of "air car," which runs on compressed air, claims to provide energy-efficient, pollution-free transportation, how could you compare this with the energy efficiency of a traditional gasoline-powered car? Do you purchase one? If you are watching your electric bill increase every year, how should you lower your monthly costs? Invest in a new heating system? Increase the insulation in your attic? Install solar panels on your roof? What are the pros and cons? To become energy-literate citizens, students will have to gain an understanding of the Five Big Ideas of energy and practice applying them to practical problems regarding energy resources. However, it takes time to develop the competencies necessary to tackle such complex practical problems.

Recent research into students' understanding of energy suggests that they can develop robust ideas about energy forms, transformations, and transfers in the elementary grades but that more sophisticated and abstract ideas, such as energy conservation and dissipation, take many more years to develop (Liu and McKeough 2005; Neumann et al. 2013). Research also suggests that students can make substantial strides toward developing a more robust understanding of energy through intentionally designed instructional experiences (Constantinou and Papadouris 2012; Nordine, Krajcik, and Fortus 2011; Trumper 1991). In the next several sections, we discuss how intentional study of energy resources can help students develop an increasingly sophisticated understanding of energy over time.

Learning About Energy Resources in Elementary School

In the elementary grades, the idea that energy can come in many forms—such as motion, light, heat, and electricity—builds an important foundation for understanding that natural energy resources exist in forms such as flowing water (motion), light from the Sun (light), and geothermal energy (heat), and that energy can be transmitted over large distances via conducting wires (electricity).

Natural resources provide both energy and materials that are critical for modern society. Elementary students should learn to recognize that they use natural resources, such as food; building materials; and specialized substances, such as metal. These natural resources can be mined or harvested from the land surface or deep in the Earth, and although many of these materials cycle naturally, they are often limited. Like natural material resources, natural energy resources are also limited, and humans can make decisions and engage in behaviors that reduce the rate at which these resources must be extracted from the Earth.

An energy perspective is useful for making sense of how the world is powered using natural resources. Like energy, natural energy resources come in different forms, each of which is connected to different forms of energy. Natural energy resources exist in forms such as materials that can be burned, sunlight, and air that is moving.

Electrical energy is not a natural resource because it does not exist in nature in an easy-to-harvest form. Yet, the vast majority of the energy that powers students' daily lives arrives in their homes as electricity. This is possible because all forms of energy can be converted into electrical energy. Many students do not ever see the coal or nuclear plants that provide them with electricity because these plants can be located far from cities and use power lines to transfer energy from the power plant system to the systems in their home.

To build an understanding that energy is critical for powering devices and that energy resources are limited, elementary students should focus on smaller-scale phenomena such as battery-powered fans or flashlights. They should notice that these devices need a battery to operate and that, over time, the battery can no longer supply sufficient energy to operate the device. A significant consideration in all engineering design challenges is determining the source and amount of energy needed to make something go. By experiencing a range of phenomena in which students can identify an energy source for a system or device and investigate how this energy resource depletes over time, students form a strong foundation for understanding the role of natural energy resources in the future.

In upper elementary grades, students should begin to recognize that all energy in animals' food was once energy from the Sun. To connect sunlight to the energy that they get from their food, students must recognize that the process of photosynthesis involves plants converting energy from sunlight into energy stored in food, and that energy in food can be transferred to other organisms when they eat plants. Furthermore, students should recognize that the energy in their food is fundamentally the same energy as that carried to plants on Earth via sunlight. These ideas contribute to a developing understanding that the energy available to society comes from either the Sun (i.e., biofuels, wind, solar, hydroelectric, coal, oil, natural gas) or Earth (i.e., geothermal, nuclear). Elementary students need to recognize that the conversion of sunlight to energy-rich food by plants is an important natural energy conversion process.

Learning About Energy Resources in Middle School

In the middle school grades, students begin to model energy transformation and transfer to track the flow of energy within and between simple systems. For example, students use arrows to track energy flow through an ecosystem as it is transferred from sunlight through different organisms. In the process, they can represent that energy is lost to the environment via thermal energy transfer (heat) each time energy is transferred between organisms; this thermal energy transfer accounts for roughly 90% of the energy that moves between trophic levels in an ecosystem. For example, keeping our bodies at 98.6°F requires food energy to be used, but no additional mass is accumulated in your body (unless you gain weight). It is important for students to be able to describe that this energy is not destroyed or used up; rather, it exists in a form that is no longer accessible to living organisms to carry on their life functions.

Students can explore loss via thermal energy transfer in simple experiments such as the cheese puff lab described in Chapter 3. As students explore transfers of energy between systems (both living and nonliving), they gain experience with the process of energy dissipation. Gathering evidence for dissipation is a crucial step toward building a deep understanding of energy conservation, as well as the conservation of energy resources. However, the idea of quantitative conservation of energy is not to be introduced at this age. Rather, conservation ideas can be implicitly introduced by helping students notice that energy transferred to a system must come from somewhere and that missing energy must have been transferred to somewhere else.

In addition to studying energy sources and transfers in living systems and food, students can use similar models and representations (e.g., arrows to indicate energy flow) to study physical systems such as electric circuits, colliding blocks, and swinging pendulums. By tracking energy transfers, transformations, and dissipation within living, nonliving, natural, and designed systems, students begin to develop a consistent set of tools for using energy to make sense of a wide range of phenomena. Over time, students can begin to recognize that the amount of energy that is available for transformation or transfer limits what is possible in a system. A pendulum cannot swing higher than the height from which it was released. The size of the battery controls how long a lightbulb can shine.[1] As students track energy transformations, transfers, and dissipation, they build an increasingly robust foundation of evidence that all systems that transform energy (whether they are technical devices or living organisms) require a steady input of energy. Maintaining such a steady input requires some naturally occurring resource that exists outside of the system.

Learning About Energy Resources in High School

In high school, students can really begin tackling more complex systems and using the law of conservation of energy to model the behavior of systems. Although conservation is perhaps the most important principle of energy, the ideas of transfer, transformation, and

---

1   A D-cell alkaline battery will generally outlast an AAA-cell alkaline battery because its larger size enables it to store more chemicals that react to produce a voltage difference between its terminals. A battery's physical size doesn't determine the voltage it provides, only the amount of reactant chemicals it can store to produce a voltage over time.

dissipation may play an even more central role when considering the use of natural energy resources. Energy is conserved overall, but energy that is easily usable (i.e., high-quality energy) is not.

As students begin to examine methods for powering systems and observe that energy sources tend to be exhausted over time (e.g., batteries run out, swinging pendulums come to rest), they begin to gain the skills needed to assess practical energy generation and delivery to communities. High school students also become more quantitative in their analysis. For example, by examining the maximum energy available within a new battery and the power consumed by an electronic device (e.g., a radio or bright lightbulb), they can predict the maximum amount of time that the device might function. Further, they can reason that any discrepancy in this value must be due to thermal losses as the energy stored in the battery is delivered to the device. Students can measure the efficiency of simple machines such as a pulley and explore what happens to the efficiency as the complexity of the pulley system increases. Likewise, students can convert units of energy so that they can compare energy efficiencies of different engines. For example, they can compare gallons of gasoline with kilowatt-hours to compare the energy demands of internal combustion versus electric engines in cars. High school students build on their understanding of energy forms, transformation, and transfer to develop their ability to use the ideas of conservation and dissipation to quantitatively model and interpret the behavior of energy-transforming systems.

Students in high school tend to begin to look outward at the larger community rather than focusing only within their own homes, and they become sensitive to political and environmental issues surrounding energy needs. They often become more engaged in solving societal problems. With this expanding perspective, students can explore how the use of fossil fuels and the associated release of carbon dioxide into the atmosphere may affect how efficiently the Earth radiates energy into space and the implications this has for Earth's climate. Students can grapple with the balance between inexpensive energy sources and environmentally sustainable ones and begin to consider the complex interplay of scientific, economic, and political factors in making environmental decisions and policies. With the knowledge that many of the Earth's natural resources (particularly fossil fuels) are limited and will someday no longer be economically feasible to use, students can examine new, innovative, renewable energy sources and assess their feasibility. In doing this sort of analysis, students make great strides toward becoming energy literate.

Teaching to Promote Understanding of Natural Energy Resources

When teaching about natural energy resources, it should be kept in mind that students are not ready to grapple with complex issues or societal needs for energy resources until high school, but that basic understandings of energy are developed within the *NGSS* for grades

K–8 so that students can approach that goal in grades 9–12. The more familiar students are with observing simple phenomena—such as electricity powering lightbulbs, radios, and microwave ovens; gasoline- or electricity-fueled cars; wind turning turbines or shafts to grind wheat; and rubber bands projecting cars down racetracks—the more ready they will be to analyze nonrenewable and renewable energy sources as they approach college. By contextualizing society's energy demands locally and empowering students to write a letter to a political leader, convince peers to invest in their energy source, and engage their classmates in debates that revolve around issues of their future standard of living, curricula focused on energy resources provide unique and powerful learning opportunities.

The following sections describe some examples for teaching about energy across K–12 with the Five Big Ideas about energy in mind.

Grades K–2

In the early grades, students should focus on the importance of light for plant growth and recognizing that light can heat materials. In addition to traditional experiments in which students explore plant growth in areas with varying light (or no light), children can also explore ways to increase or decrease the amount that an object heats up when exposed to light. For example, children could go outside on a sunny day and be given a set of materials (e.g., straws, different colors/thickness of paper, tape) to use to build something that decreases the rate at which an ice cube melts on a tray. In this activity, students should recognize that the more protected they make the ice cube from the sunlight, the more slowly it will melt. Students at this age need not connect this activity to the term *energy* or even to make measurements of temperature. The most important thing is that they recognize that the lower level of sunlight hitting the ice cube, the more slowly it will melt. This connection sets the stage for more sophisticated investigations later in elementary school.

Grades 3–5

In upper elementary grades, students should focus their learning on observations of phenomena that highlight different energy forms, sources, and transformations. For example, the FOSS (Full Option Science System) Energy and Electromagnetism module (grades 4–6) allows students to build test circuits and test different objects for conductivity.[2] They can also manipulate their test circuits to explore series and parallel circuits and determine the pathway(s) for current to flow in a circuit. Snap Circuits are excellent resources to extend these electricity explorations.

Safety Notes
1. Wear sanitized safety glasses or goggles when working with wires having exposed metal—sharp hazard!
2. Stay clear of water or other liquids when working with live wires—shock hazard!

---

2 FOSS Third Edition: Energy and Electromagnetism module, 2012. Developed at the Lawrence Hall of Science and published and distributed by School Specialty Science/Delta Education. Copyright © The Regents of the University of California.

Colorful components and connectors can be snapped together to form any number of circuits. In all cases, students quickly discover that an energy source is needed: a solar panel, a wind turbine, or a battery needs to be hooked up to make things go. In addition, the solar panel needs a source of light, and the wind turbine needs moving air for anything to happen. Students learn that energy comes in different forms but is necessary for all circuits.

Students also want to design circuits to do something. They can add a lightbulb, a speaker, a gear motor, or a blower. Then, if they connect the circuit completely and correctly, they see the electrical energy converted to another form: sound, light, or movement. Although they do not have to talk explicitly about energy being transferred, conserved, or dissipated, their explorations give them strong experiences that demonstrate that the energy source can run out (e.g., a dead battery, too little light, too little wind) and more energy is required to last longer, make louder sounds, or move larger objects. By focusing on phenomena, students can gain experience with different forms of energy (e.g., electricity, sunlight, wind) that can be transferred from one system to another to make things happen (e.g., light a bulb, make music). They do not yet have to worry about conservation of energy or dissipation, just that energy is required to do things and there are different kinds of energy in different situations.

Grades 6–8

At the middle school level, students are ready to apply the ideas that energy can take on different forms and be transferred from one system to another. To apply these ideas, students can consider the consequences of powering systems such as homes and cars with different energy sources. This allows students to evaluate the benefits and drawbacks of using different energy resources and be introduced to ideas of energy efficiency, energy losses, and renewable energy resources. Although students have not yet developed a fully quantitative approach to using energy to analyze complex systems, they can be engaged through data and discussions of local energy use and how it has changed in the past and may change in the future.

Joy Reynolds, who teaches in the Detroit Public Schools, teaches a unit on energy resources in Michigan to engage her middle school students in learning about the energy resources that power their state. This unit is supported by multiple materials from the Michigan Environmental Education Curriculum Support (MEECS) project (Michigan Department of Environmental Quality n.d.), a seven-unit curriculum focused on learning about Michigan's economy and environment provided by the Michigan Department of Environmental Quality. The MEECS "Energy Resources" unit is composed of eight lessons covering electricity generation, renewable and nonrenewable energy resources, energy conservation, and sustainability.

In the first lesson, "Energy Use in Michigan, Then and Now," students consider the following essential questions: (1) How do we use energy in Michigan? (2) How and why has our energy consumption changed over the last 100 years? and (3) How does Michigan's energy use compare with that of the whole United States? Students brainstorm and categorize the uses of energy, interpret graphs about energy use, take surveys, and engage in small-group and classroom discussions about energy comparisons. In the next seven lessons of the unit, students examine Michigan's energy resource mix, build a simple turbine to investigate the process of electricity generation, explore nonrenewable energy choices and impacts, explore advantages and disadvantages of renewable energy resources, consider strategies for energy conservation and increasing energy efficiency, model a product life cycle, and strategize how to leave a smaller ecological footprint.

During these lessons, students engage in scientific practices (e.g., modeling, defining problems, analyzing data, engaging in argument, and communicating information), build disciplinary core ideas (e.g., Conservation of Energy; Energy Transfer and Energy in Chemical Processes in Everyday Life), and use crosscutting concepts (e.g., Systems and System Models; Energy and Matter: Flows, Cycles, and Conservation) while engaging in a study with a local context. By investigating energy resource use in the region where they live, students investigate an authentic and relevant problem that promotes three-dimensional learning.

Grades 9–12

At the high school level, students can engage in more sophisticated analysis of systems and begin to use all Five Big Ideas about energy using quantitative analytical techniques.

Amanda Chapman engages her physics students in lessons that, like Joy Reynolds' lessons in Michigan, explore the benefits and drawbacks of using different energy resources to power their local context—in this case, the city of San Antonio, Texas. Amanda has designed her physics lessons to address the work-energy relationship and the concept of power. Students launch cars with rubber bands and use a photogate and spring scale to measure the speed of the cart as it is launched and the average force it takes to pull back the rubber band a certain distance. Using these data, students calculate the potential energy of the stretched rubber band and the kinetic energy of the car as it is launched. Rather than ignore the small difference between elastic potential energy and kinetic energy as experimental error, Amanda highlights the fact that the elastic potential energy is always just a little bit greater than the kinetic energy that they measure. This forms the foundation for exploring energy dissipation and the efficiency of machines.

Safety Note
Wear sanitized safety glasses or goggles when doing these activities.

During the 16 lessons (each 55 minutes), students conduct experiments, solve conservation of energy problems, calculate spring constants, and explore the rate at which energy is transferred via mechanical work when walking or running up the stairs (power). The final three lessons build on students' experiences with work-energy-power relationships and apply them to address the original question of what energy sources San Antonio should use. The unit culminates in students writing a letter to the head of a local energy council advising what source of energy the city should use and why. By studying energy within a real-world energy resource context, students are in a position to apply the concepts of energy transformation in devices, transfer between objects and systems, and energy conservation and dissipation in ways that are likely to help them see connections between traditional physics experiments and their local context.

Another example of high school teaching of energy resources comes from Huang Yanning, who teaches an 11th grade chemistry course for honor students in China. In the unit called "The Heat of Combustion of Fuels—The Comprehensive Utilization of Coal," these students examine Hess's law (i.e., the idea that the enthalpy of a chemical reaction is a state function; it is independent of the path or reaction steps taken from the initial to final state), the heat of combustion, and enthalpy change through an examination of the major fossil fuel in China—coal. Students are challenged to select a fuel for optimum use by examining the enthalpy changes of combustion reactions of various fuels, such as natural gas, coal, ethanol, and hydrogen. They are challenged to address questions such as, "Why do urban families mostly use natural gas?" "Why should we make sure that fuels burn completely to carbon dioxide?" "How do various fuels affect the environment?" "What is the advantage of coal gasification (converting coal to natural gas)?"

By using the context of energy resources, this unit on chemical reactions and Hess's law[3] comes alive. Students explicitly examine different forms or sources of energy and implicitly consider that each of these sources is converted to the electricity needed in their homes. They study the chemical reactions that allow energy to be transferred from chemical potential energy to heat (i.e., enthalpy) that can be used to generate electricity. Students thus apply their knowledge of the Five Big Ideas of energy to a large issue of societal importance. This reinforces prior learning about different forms of energy, the transformation of energy from natural resources to forms that can be used by society, the need to transfer energy from natural to human-designed systems while trying to minimize dissipation into the environment, and that energy is conserved on the microscopic (molecular bonds) level as well as quantitatively on the macroscopic level.

Teaching Natural Energy Resources Over Time: The Water Cycle Example

Learning about energy resources can be integrated into any grade level, but teachers need to map the Five Big Ideas of energy with the *NGSS* performance assessments and their available curricula. As an illustration, the water cycle (i.e., the process of the movement of water through various pools on the surface of the Earth through inputs of energy driving

---

3 Hess's law is a chemical representation of the law of conservation in that energy can be accounted for in all chemical reaction pathways, but some energy is consumed or released in the conversion of the initial state of chemical reactants to the final state of chemical products.

processes such as evaporation, condensation, and precipitation) can be taught in different grade bands using different emphases and different fundamental concepts to lead to an in-depth analysis of hydroelectric power by the end of high school.

In elementary grades, students should notice that water always flows downhill and that water moves and can make things move. Students should learn that water can become ice when it cools or steam when it is heated. They can explore the effects of light and heat on the melting of ice or evaporation of water. By exploring fundamental components of the water cycle from a phenomenological perspective and connecting these components to energy transfer mechanisms such as heat and light, students can develop a strong foundation from which to be introduced to the water cycle.

In the middle grades, students can observe water being evaporated and condensed in a terrarium and extend their thinking about this system to create diagrams representing the water cycle on Earth. Students can also use stream tables and topological models to study watersheds that have rain and groundwater flowing into tributaries, into the main stem rivers, and then down to the ocean. By connecting this movement and pooling of water to changes in gravitational energy, students can recognize that gravitational potential energy decreases as water flows downhill and that there must be an energy input for this same water to rise in the atmosphere; the Sun provides this energy.

In high school, students can explore energy transfers and transformations during phase changes and use an atomic-molecular perspective to explain why an energy input is needed to ensure that the water cycle continues over time. By analyzing the gravitational potential energy that exists in different reservoirs and its process of transformation into electric energy that can be transferred along power lines, students can explore why dams are necessary to efficiently generate electricity from water flowing in a river. A thorough understanding of energy and the water cycle might also help students discover that energy can be efficiently stored during the day by pumping water up to fill a reservoir so that the same water can generate hydroelectric power at night (when solar panels cannot supply needed energy for lighting our homes).[4]

Learning about the water cycle means much more than naming its parts and drawing arrows on a picture of the ocean, land, and clouds. As students build their understanding of energy over time, they can begin to understand and interpret the water cycle as a system that relies on energy transformation and transfers to account for the fact that energy is dissipated as the cycle continues.

---

4 This is a clever idea that is one of the promising solutions for supplying uniform electricity when renewable resources such as sunlight can only supply electricity during one part of the day. This is much more efficient than storing energy in a battery.

Summary

Natural resources are used in society to increase our standard of living. Energy is one of the most important natural resources that the Earth provides for us, but forms of available and usable, high-quality energy resources (i.e., those that are easy to use) are limited. Energy can be extracted from fossil fuels (i.e., coal, oil, natural gas), wind, sunlight, geothermal gradients, moving water, nuclear fuels, or biofuels. Each of these natural energy sources requires designed systems to extract (i.e., transfer from the natural resource system), convert, and deliver energy to end users (i.e., transfer to end users), and each of these processes has a different efficiency (i.e., a different amount of energy is dissipated as thermal energy in the process). Through targeted instruction beginning in elementary school and lasting throughout high school, we can help students apply the Five Big Ideas of energy to systems that span the natural and designed world to understand how we use natural resources to power our day-to-day lives. Recognizing that all energy is fundamentally the same but can be manifested in different forms will help students explain what changes are necessary to use natural energy resources into the future and why many resources can be combined to provide the energy we need. By tracking energy transformations and transfers within and between systems, students learn that dissipation is a fact of life. Because energy is conserved, a continuous input of energy is required to sustain a resource from which we are extracting energy; the balance of input to extraction determines whether resources are renewable or not.

Natural energy resources are an excellent topic to use when exploring energy as a crosscutting idea, because the process of evaluating and using natural energy resources involves challenges that are biological, geological, chemical, and physical; designing systems to use natural energy resources is an engineering problem with tremendous social consequence. With an understanding of the Five Big Ideas of energy and practice applying them to authentic problems related to energy resources, students are well positioned to develop the kind of energy literacy that is critical for tomorrow's citizens and scientists as they confront the grand challenge of powering our society in a sustainable way.

References

Boyes, E., and M. Stanisstreet. 1990. Misunderstandings of "law" and "conservation": A study of pupils' meanings for these terms. *School Science Review* 72: 51–57.

Constantinou, C. P., and N. Papadouris. 2012. Teaching and learning about energy in middle school: An argument for an epistemic approach. *Studies in Science Education* 48 (2): 161–186. *http://doi.org/10.1080/03057267.2012.726528.*

Intergovernmental Panel on Climate Change. 2000. *Special report on emissions scenarios: A special report of Working Group III of the Intergovernmental Panel on Climate Change,* eds. N. Nakićenović and R. Swart. Cambridge, England: Cambridge University Press.

Liu, X., and A. McKeough. 2005. Developmental growth in students' concept of energy: Analysis of selected items from the TIMSS database. *Journal of Research in Science Teaching* 42 (5): 493–517.

Michigan Department of Environmental Quality. 2015. What is MEECS? *www.michigan.gov/deq/0,1607,7-135-3307_3580_29678-148152--,00.html.*

National Research Council (NRC). 2012. *A framework for K–12 science education: Practices, crosscutting concepts, and core ideas.* Washington, DC: National Academies Press.

Neumann, K., T. Viering, W. J. Boone, and H. E. Fischer. 2013. Towards a learning progression of energy. *Journal of Research in Science Teaching* 50 (2): 162–188. Also available online at *http://doi.org/10.1002/tea.21061.*

NGSS Lead States. 2013. *Next Generation Science Standards: For states, by states.* Washington, DC: National Academies Press. *www.nextgenscience.org/next-generation-science-standards.*

Nordine, J., J. Krajcik, D. Fortus. 2011. Transforming energy instruction in middle school to support integrated understanding and future learning. *Science Education* 95 (4): 670–699. Also available online at *http://doi.org/10.1002/sce.20423.*

Trumper, R. 1991. Being constructive: An alternative approach to teaching the energy concept—part two. *International Journal of Science Education* 13: 1–10.

U.S. Department of Energy. 2014. *Energy literacy: Essential principles and fundamental concepts for energy education.* Washington, DC: U.S. Department of Energy. *http://energy.gov/sites/prod/files/2014/09/f18/Energy_Literacy_High_Res_3.0.pdf.*

U.S. Energy Information Administration. 2013. *International energy outlook 2013.* No. DOE/EIA-0484(2013). Washington, DC: U.S. Energy Information Administration.

CHAPTER 8

ASSESSING ENERGY AS A CROSSCUTTING CONCEPT

KNUT NEUMANN, DAVID FORTUS, AND JEFFREY NORDINE

A Car Conundrum

A car dealer shows a customer a fantastic van on sale. It provides plenty of space to carry a family of four and their entire luggage. It is very sturdy because it has steel reinforcements all around and even specially reinforced tires. Despite all the weight these features add to the van (it weighs 2,500 pounds), the van—says the dealer—gets about 100 miles per gallon of gasoline when traveling on the highway. Best of all, it comes with a price tag of just $9,999! What should our customer say?

As science teachers, we might hope that our customer would run screaming from the dealership. We hope that our customer understands that accelerating the mass of such a car (not to mention the family of four and their belongings) and keeping it moving at highway speeds despite the air resistance (i.e., drag) of its bulky shape and rolling resistance of its reinforced tires will require plenty of energy, definitely much more than a compact car. Given the efficiency of any (currently known) combustion engine, the energy released from burning a gallon of gas will simply not provide enough energy to move the van anything close to 100 miles.

Having left the unscrupulous dealership, our customer starts looking for alternatives and begins considering an electric car. The car dealer selling this car says that although the car is expensive ($45,000 MSRP), it ends up being a bargain because it comes with a new kinetic energy recovery system. This system kicks in when you brake, and if the driver brakes gently enough, it is possible for the system to recover 100% of the kinetic energy of the moving car and return it to the electric battery. Even if your braking behavior is not sufficiently gentle, there is no need to feel bad, because the car runs on clean energy anyway. What would we hope our customer says to this dealer?

As science teachers, we would hope our customer understands that it is impossible to recover all the energy in the braking process (because although energy is indeed conserved overall, there is no rule that it stays in the same system—some will inevitably be lost to thermal energy and transferred to the surroundings; see Chapter 2). We would also hope that our customer understands that while people may say that electric cars run on clean energy because they do not directly create any pollution, the electrical energy required to run these cars is obtained through the use of natural energy resources—and, in the United States, this usually means the burning of coal in a coal-fired power plant (see Chapter 7).

A deep understanding of the energy concept helps students know when things seem too good to be true and what questions to ask when making energy-related decisions. Using energy ideas in such cases involves much more than just being able to recite the fact that energy is never created or destroyed. The processes of energy dissipation, transfer, and transformation are critical when considering factors such as efficiency and the environmental impact of automobiles. Beyond deciding which car is best for them, we want students to be prepared to make decisions about other important issues. We want students to become actively involved with issues that have societal relevance, such as whether the United States should phase out nuclear energy like the Germans did after the incident in Fukushima, Japan, or what to do about global warming.

Assessing Energy Understanding

One important outcome of K–12 science education is enabling students to analyze and develop possible solutions for some of the complex, real-world problems they may encounter in their later lives. Although it is these kinds of problems we want students to be prepared for, they are often not the kind of problems we present our students in school. Instead, when it comes to assessing what students have learned in school about a concept such as energy, the tasks often focus on problems in simple contexts that fail to illuminate whether students have developed a deep understanding that is useful for approaching real-world problems. Consider, for example, the assessment item shown in Figure 8.1.

If you teach biology, you may have seen similar assessment items or even included something like it on a test or quiz. In many ways, this is not a bad item: it involves a familiar context, there is one clear answer and three clearly wrong answers, each option is plausible, and the idea tested is important in biology. The item targets the idea that from one trophic level to the next, roughly 10% of the energy associated with the lower trophic level is still present within the ecosystem (i.e., the 10% rule). (To choose the correct answer, students need to recognize that plants and birds are two trophic levels apart. So, start with 100%, take 10% for transfer from one trophic level to the immediate next level, and then take 10% again for transfer to the next level—resulting in 1% of the original amount.) On the surface, this item seems to align with students' understanding of energy transfer and dissipation because it involves energy moving between trophic levels (energy transfer)

Figure 8.1. An assessment item asking students about energy flow in an ecosystem

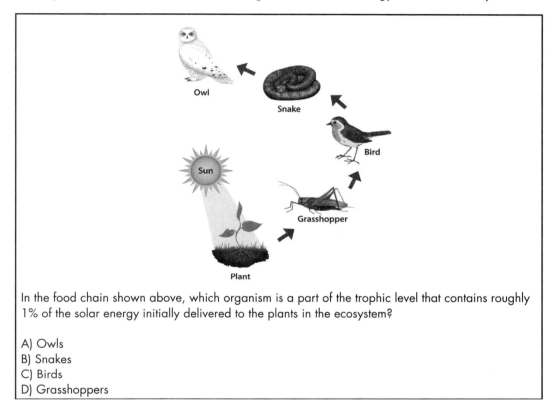

In the food chain shown above, which organism is a part of the trophic level that contains roughly 1% of the solar energy initially delivered to the plants in the ecosystem?

A) Owls
B) Snakes
C) Birds
D) Grasshoppers

and the loss of 90% of energy between trophic levels (dissipation). But a closer examination brings into question whether students really must know and use these ideas to answer the question correctly.

While it is a fine item in many ways, the item in Figure 8.1 fails to require students to demonstrate an understanding of key aspects of energy transfer and dissipation. If a student knows the 10% rule, this is often enough to answer the question. Students needn't consider the system involved (i.e., that the 10% rule applies to trophic levels and not to individual organisms), the form of the energy (i.e., that the energy contained within an ecosystem is that which, in the form of chemical energy, is made accessible through digestion), or why 90% of the energy from the previous trophic level is suddenly gone (i.e., students needn't identify where the rest of the energy has gone). This item provides teachers with little or no information regarding whether a student knows how the 10% rule is connected to any of the Five Big Ideas about energy (see Chapter 1).

Figure 8.2 (p. 142) shows an example of an item that is worded in a way that is more tuned to the Five Big Ideas of energy. This item was part of an end-of-year summative

Figure 8.2. Skater in a half-pipe item

A skater goes back and forth in a half-pipe without pushing on the half-pipe with her feet.
How can skateboarding be described using energy?
A) The kinetic energy of the skater is converted into gravitational energy when the skater is moving up the walls of the half-pipe. This gravitational energy is converted back into kinetic energy while the skater is moving down the walls of the half-pipe.
B) Gravitational energy of the skater is converted into kinetic energy while the skater is moving up the walls of the half-pipe. This kinetic energy of the skater is converted when the skater is moving down the walls of the half-pipe.
C) The skater gets power out of chemical energy of his food. The skater can convert this power into kinetic and gravitational energy without having to push.
D) The gravitational energy of the skater is converted into kinetic energy while the skater is moving down the walls of the half-pipe. As a result, the skater gets power to go up the walls on the other side of the half-pipe.

Source: Adapted from Neumann et al. 2013. Used with permission.

assessment to monitor students' progress in understanding energy (Neumann et al. 2013). Compared with the item in Figure 8.1, this item is more specifically tuned to one of the Five Big Ideas of energy: that one form of energy can be converted to other different forms. To correctly answer this item, students must be able to associate forms of energy with the processes that convert one into another when a skater rides on a half-pipe.

Both of the items in Figures 8.1 and 8.2 can tell us if students remember and, depending on whether we have taught them about the specific context of the items, apply what we have specifically taught them during instruction. However, they provide little information about whether students can apply the individual Big Ideas about energy together (i.e., connect them) to solve tasks that confront them with authentic real-world problems.

Assessing Energy as a Crosscutting Concept

Authentic real-world problems are not necessarily confined to individual disciplinary contexts. In fact, most real-world problems span multiple disciplines. These problems require more than the knowledge about the individual Big Ideas; they require students to connect and apply them across different disciplinary contexts. That is, real-world problems require an understanding of energy not only as a disciplinary idea but as a crosscutting concept (CCC), and the ability to apply this understanding to make sense of authentic scenarios.

Take, for example, the demonstration of a student holding a heavy book described in Chapter 4. The student volunteer holds a book in her outstretched arm and the teacher triumphantly declares that the student has done no work in this process because the book does not move. But then, why does the student get tired after some time if she does no work? The demonstration emphasizes a very physics-specific way of thinking about work (or energy transfer via force), and one that often does not map well onto biological contexts because it fails to address the question of why the student gets tired. This task can do better to emphasize the crosscutting nature of energy if, instead of asking whether work is done, the teacher asks students to consider evidence of energy transformations and transfers. Students should consider that over time, the student will get hot and begin to sweat and that this is evidence of energy transformation and, eventually, transfer to the surroundings. In doing so, students can also consider whether energy is transferred to the book and recognize that it is not (if the book is assumed perfectly still). In this more crosscutting perspective, students both consider the work done (or not done) on the book and also address the question of why the student gets tired. Such analysis addresses and helps connect energy as it is often used in physical and biochemical contexts and elicits information about whether students are connecting ideas across contexts within a real-world situation.

Similarly, we can use the skater in the half-pipe item (Figure 8.2) to assess students' understanding of energy in both physical and biochemical contexts. If we prompt students to go beyond simply describing how gravitational energy is converted into kinetic energy (and vice versa) and also ask them why the skater will eventually end up standing still at the bottom of the half-pipe, we require students to invoke multiple Big Ideas about energy. While the skater is going up and down in the half-pipe, some of her kinetic energy is being converted into thermal energy that spreads out across the skateboard, half-pipe, skater, and surrounding air. Thus, if the skater does not push on the half-pipe, she will eventually come to a rest. Additionally, the skater needs to eat (and breathe) to provide the energy input needed to push on the half-pipe. Most of this energy released during cellular respiration actually goes into running metabolic processes within the skater's body—only a small part of this energy released during cellular respiration is then converted into the kinetic and gravitational energy of the skater as she rides. Nearly all of the released energy is ultimately converted to thermal energy in the surroundings as she stops riding.

The book-holding demonstration and the skater in a half-pipe scenario are both examples of real-world contexts that can be used to elicit students' understanding of the Five Big Ideas about energy. It is important to note, however, that assessing (and developing) an understanding of energy as a CCC requires more than a simple tweak of existing disciplinary assessments and activities; it requires providing students with the opportunity to connect multiple ideas about energy across disciplinary contexts to solve specific problems.

This, in turn, requires well-chosen real-world contexts that span across multiple disciplines, carefully designed tasks confronting students with a specific problem to solve, and a scoring scheme that provides us with the information about where students reside on their way to developing an understanding of energy as a CCC. Also, to foster such understanding, our students need to become aware that we expect more from them than simply remembering single ideas about energy and applying them to disciplinary problems. This expectation must be conveyed in instruction and repeatedly reinforced in assessments.

The *Next Generation Science Standards* (*NGSS;* NGSS Lead States 2013) and *A Framework for K–12 Science Education* (NRC 2012) provide us with a powerful set of recommendations for promoting and assessing richer understandings of science and scientific ideas—the vision of three-dimensional learning. In this vision, students are expected to go beyond simply remembering and applying a set of ideas from selected science disciplines (i.e., disciplinary content knowledge). Instead, students are expected to develop the competence to apply a wealth of scientific practices, together with their content knowledge about disciplinary core ideas (DCIs), to solve problems cutting across disciplinary boundaries using CCCs. Energy is both a DCI and a CCC, and the Five Big Ideas are the bridge between using energy as a disciplinary idea and as a crosscutting concept. But students also need to use scientific practices as they build and demonstrate understanding of energy, and the *Framework* reminds us that scientific practices are developed over time. Assessments need to target students' ability to apply different scientific practices together with their understanding of energy to solve problems that require a cross-disciplinary understanding of energy.

Organization of This Chapter

In the following sections, we consider different examples of how assessment tasks can embody the vision of the *Framework* and focus on the Five Big Ideas of energy. To do so, we will focus on the three learning goals expressed in the *Framework* that guided the summit on the teaching and learning of energy (which formed the foundation for this book) and that emphasize different aspects of the crosscutting nature of energy (see Chapters 5, 6, and 7):

1. Many organisms use the energy from light to make sugars (food) from carbon dioxide from the atmosphere and water through the process of photosynthesis, which also releases oxygen. In most animals and plants, oxygen reacts with carbon-containing molecules (sugars) to provide energy and produce waste carbon dioxide. (LS1.C)

2. That there is a single quantity called energy is due to the remarkable fact that a system's total energy is conserved as smaller quantities of energy are transferred between subsystems—or into and out of the system through diverse mechanisms and stored in various ways. (PS3.A)

3. All materials, energy, and fuels that humans use are derived from natural sources, and their use affects the environment in multiple ways. Some resources are renewable over time, and others are not. (ESS3.A)

Each of these three learning goals includes a cross-disciplinary perspective that needs to be systematically developed throughout grades K–12. To discuss how we can assess these learning goals across disciplines and at different grades without presenting an overwhelming number of examples, we will consider one learning goal across disciplines at the same grade band, one learning goal across grade bands in the same discipline, and one learning goal an example of blending ideas from different disciplines into assessment contexts.

In the following sections, we begin with the second learning goal because it cuts to the core of what energy is as a scientific concept. In a series of examples using a Rube Goldberg machine, we discuss how middle school students' progress toward this learning goal can be assessed across multiple disciplines. We focus on middle school because it represents an intermediate level of understanding and because middle school contexts can most readily be adapted for use in elementary school or high school. Then, returning to the first learning goal, we will illustrate how students' developing understanding of the role of energy in photosynthesis and cellular respiration can be assessed across elementary, middle, and high school. Next, we will explore how the third learning goal (which relates to natural energy resources) can be assessed across grades by synthesizing energy ideas from biology, chemistry, Earth science, and physics. After discussing examples from each learning goal, we will zoom out to discuss broader assessment principles that align with the *NGSS* and consider guidelines for developing assessments of three-dimensional learning (NRC 2014) in light of these principles.

Assessing Students' Understanding of Energy Conservation

The *Framework* (NRC 2012) describes the idea of energy conservation in this way:

> *That there is a single quantity called energy is due to the remarkable fact that a system's total energy is conserved as smaller quantities of energy are transferred between subsystems—or into and out of the system through diverse mechanisms and stored in various ways. (p. 123)*

How can we assess whether students have made adequate progress toward this learning goal? That is, what would a task to assess students' learning with respect to this goal look like? What activities might you use in your classroom to engage students in learning about energy conservation?

You may have asked students to explore a swinging pendulum, a mass bobbing on a spring, or a roller coaster. But often we ask students to assume that energy is conserved for these simple systems, which it is not because energy is transferred to the surroundings as they move. Asking students to make such an assumption not only fails to reflect reality, it also misses the opportunity to emphasize the second part of the *Framework*'s statement about energy conservation—that the principle is perhaps best understood in the context of analyzing how energy is transferred between systems and subsystems through a variety of mechanisms. Rather than asking students to focus on idealized systems, what if we assessed their understanding of energy conservation through a real system that transfers and stores energy in a variety of mechanisms? Enter the Rube Goldberg machine, which is useful for assessing energy conservation in a physics context.

Assessing Energy Conservation in a Physics Context: The Rube Goldberg Machine

A Rube Goldberg machine has no practical use. It is a fun device that completes a usually somewhat simple task in an over-engineered, very complicated fashion. To do so, the machine is usually designed as a chain or sequence of smaller devices designed to complete even simpler, mostly meaningless tasks, such as a ball rolling down a track to hit a mousetrap, which sparks a lighter to light a candle, which heats water, and so on. Famously elaborate examples of Rube Goldberg machines are shown in a commercial by Honda called "The Cog" and a music video by the band OK Go for the song "This Too Shall Pass" (both videos are easily found on virtually any internet video site—e.g., *www.youtube.com/watch?v=_ve4M4UsJQo* and *www.youtube.com/watch?v=qybUFnY7Y8w*).

Rube Goldberg machines can make great activities for both instruction and assessment. For example, middle school students can design and build a Rube Goldberg machine and then be asked to tell the story of their machine and represent the energy transfers and transformations that take place as it operates. Figure 8.3 shows an example of how students might represent the story of a simple Rube Goldberg machine.

> **Safety Note**
> Wear sanitized safety glasses or goggles during pre-lab setup, the activity, and post-lab cleanup.

> **TASK:** Rube Goldberg machine
> **GRADE BAND:** Middle school
> **DCIs:** Energy (PS3.A, PS3.B); Engineering design (ETS1.A, ETS1.B)
> **CCC:** Energy and matter: Flows, cycles, and conservation
> **PRACTICE:** Developing and using models

After students tell the story of how their Rube Goldberg machine operates, they can construct a representation of the energy story of their contraption. Figure 8.4 (p. 148) shows an example of how students might represent the energy story of the machine in Figure 8.3; they can represent energy transformations (from one form to another, Big Ideas 1 and 2) and transfers (from one subsystem or object to another, Big Idea 3) that take place as their contraption operates.

Figure 8.3. Sample Rube Goldberg machine

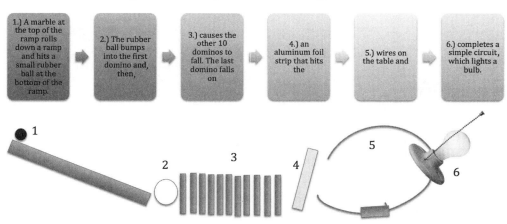

Source: Ann Novak, Greenhills School. Used with permission.

In this contraption, the marble begins with gravitational energy (because it is located at a certain height above the ground) that is converted into kinetic energy as the marble moves down the ramp, losing height and gaining speed. The amount of kinetic energy is smaller than the gravitational energy the marble had due to its height above the ground, but this energy still exists (conservation, Big Idea 4) as thermal energy of the apparatus and the environment as the ball moves down. This thermal energy spreads out in the environment and is no longer available for powering the contraption (dissipation, Big Idea 5). As the machine continues to operate, the remaining kinetic energy initiates the remaining series of steps. Notice that throughout Figure 8.4 (p. 148), the representation includes the evidence that each energy transfer and transformation is taking place.

In a Rube Goldberg machine activity like this, students can demonstrate their understanding of each of the Five Big Ideas of energy and the connections between them by constructing a model that represents the energy story in a real, nonidealized apparatus that they created. This activity can assess students' learning about energy as a core idea, the scientific practice of modeling, and energy as a CCC.

Depending on the level of your students' interest (or engineering abilities), you may choose to provide students with the materials to be used to build more complex machines. Your students may be interested in including robotic elements, or you may choose to let students bring in bits and pieces of material they acquire themselves. You may differentiate the activity by providing students with a base set of materials and asking them to add parts to the machine. Students can describe their artifacts using different representations. They may use representations similar to the one shown in Figure 8.4 (p. 148) to represent energy transformations and transfers. If you want students to be more specific about of the types and relative amounts of energy available in the different stages, you may ask them to use

Figure 8.4. The energy story of the Rube Goldberg machine pictured in Figure 8.3

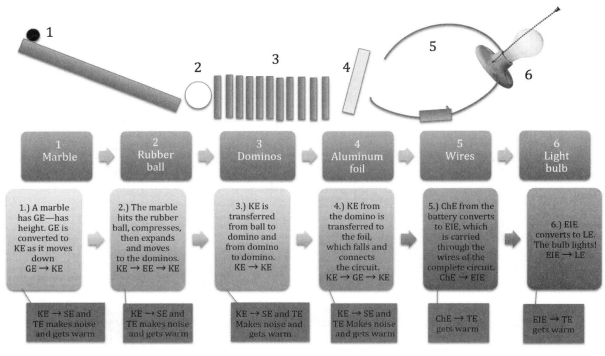

Note: ChE = chemical energy; EE = elastic energy; ElE = electrical energy; GE = gravitational energy; KE = kinetic energy; LE = light energy; SE = sound energy; TE = thermal energy.

Source: Ann Novak, Greenhills School. Used with permission.

different representations (e.g., pie charts that emphasize how the quantity of energy in the Rube Goldberg system changes over time).

The models constructed by students can help you determine where they are in their progress toward meeting the learning goal. Are students able to correctly identify the forms involved? Can they correctly identify the amount of energy available in different forms? Can they link the idea of energy transfer (i.e., energy stored in a particular way being moved around) and transformation (i.e., energy being stored in a particular way being re-stored in a different way)? Do they account for the fact that in each step (i.e., subpart) of the Rube Goldberg machine some of the energy is converted into thermal energy? Does the total amount of

---

TASK: Rube Goldberg machine
GRADE BAND: Elementary school
DCIs: Energy (PS3.A, PS3.B); Engineering design (ETS1.A, ETS1.B)
CCC: Energy and matter: Flows, cycles, and conservation
PRACTICE: Developing and using models

TASK: Rube Goldberg machine
GRADE BAND: High school
DCIs: Energy (PS3.A, PS3.B)' Engineering design (ETS1.A, ETS1.B)
CCC: Energy and matter: Flows, cycles, and conservation
PRACTICE: Developing and using models

---

energy in various forms and various places in each step (i.e., subpart of the machine) add up to the total amount of energy in the preceding and following part of the machine? Middle school students who are able to fully and correctly model the energy processes going on in the machine they have built, linking the different Big Ideas about energy together, have demonstrated strong evidence that they are developing age-appropriate knowledge that is important for developing a robust understanding of energy conservation over time.

Although we are focusing on using the Rube Goldberg machine at the middle school level, it is certainly not an inherently middle school context and can be useful for assessing a wide range of developmental stages. Elementary students can build simpler machines and models, and high school students can quantitatively model more complex machines that include a wider array of energy transformations and transfers. Rube Goldberg machines are also an excellent way to connect to engineering principles, such as precisely defining the criteria and constraints of a design task (ETS1.A) and testing and improving design solutions (ETS1.B).

The Rube Goldberg machine activity is just one example of a real-world context that can be used to assess whether students have met our learning goal. Although it reinforces the Five Big Ideas about energy, it is an activity that largely emphasizes energy in a physics context. To assess students' crosscutting understanding of energy, we need other tasks that can emphasize the same ideas about energy but in other disciplinary contexts.

Assessing Energy Conservation in a Biology Context: Priestley's Experiment

In a biology context, students can be introduced to a famous experiment conducted by Joseph Priestley. Priestley lived in the 18th century—a time when it was well known that animals could not survive in a sealed glass container. Priestley noticed that relative to animals, plants could survive much longer in those containers. To further explore this phenomenon, he put one mouse in a sealed glass container and another mouse in another glass container, but this time with a plant. The plant was separated from the mouse such that matter could flow between the two parts of the container but the mouse could not eat the plant. Priestley observed that the mouse in the container with the plant stayed alive longer than the other mouse.

Priestley's experiment involves the flow of matter, as chemical reactions occur within the plant (through photosynthesis and cellular respiration) and the mouse (through cellular respiration). Further, energy carried by light from the Sun is transferred to the plant and used to construct glucose molecules during photosynthesis. During cellular respiration, the energy used to construct the glucose molecules is

> **TASK:** Priestley experiment
> **GRADE BAND:** Middle school
> **DCIs:** Energy (PS3.A, PS3.B, PS3.D); From molecules to organisms: Structures and processes (LS1.C)
> **CCC:** Energy and matter: Flows, cycles, and conservation
> **PRACTICE:** Developing and using models

released and becomes available for carrying on life functions. As a consequence, Priestley's experiment can be an excellent opportunity to gauge students' understanding of both matter and energy flows in living systems.

The activity shown in Figure 8.5 is designed to provide students with the opportunity to demonstrate their understanding of the matter and energy flows in the context of Priestley's experiment. The task requires five teams of students. Each group is assigned one organism of a food web (e.g., algae, aquatic plant, aquatic insect, mouse, or fox). The first two items of the task require students to write down which organism they were assigned and whether the organism performs cellular respiration, photosynthesis, or both. Items 3–5 aim to activate students' prior knowledge by having them identify the flow of matter into and out of the specific organism of the food web assigned to them. Next, students identify the energy forms that their organism can use and are prompted to think about how these energy forms are converted (or not) when energy is transferred to or from organism. This task requires students to develop a model of the energy transfer and transformation through a simplified ecosystem.

The Priestley experiment task assesses very similar ideas as the Rube Goldberg machine task, except that in the Priestley experiment task students are responsible for describing only individual parts of a larger system. By bringing the whole class together at the end of the activity or, perhaps, by using a jigsaw cooperative grouping strategy (Aronson and Patnoe 1997) in which new groups are formed that contain at least one representative from each of the first groups, students can put their work together to describe the flow of matter and energy through the entire ecosystem under study.

Assessing student performance on the Priestley experiment task can be done with the Five Big Ideas about energy in mind. Do students, for example, simply name different forms of energy in their responses, or do they indicate that all forms are fundamentally the same (Big Idea 1)? Do they identify all the energy transformations (Big Idea 2) or transfer processes (Big Idea 3) going on? Do they recognize that when energy passes through an organism, although some of it appears to be gone, it still exists as thermal energy (Big Idea 4) that dissipated and so is no longer useful for driving life processes (Big Idea 5)?

If we want to see whether students have an (at least qualitative) understanding of the conservation of energy, we could also ask them to quantify the relative amounts of energy involved using pie charts. By comparing students' descriptions of energy flow in individual organisms and in the entire food web, we can assess whether students seem to be consistently applying the ideas of energy forms, transformations, transfers, conservation, and dissipation.

Figure 8.5. Activity sheet on the flows of energy and matter through a food web (sample answers in parentheses)

Purpose

Investigate the following question: How does the flow of energy relate to the flow of matter between organisms in an ecosystem?

Procedure

Answer the following questions about the organism you are using to investigate the flow of matter and energy.

1. Write the name of organism you are working with in the middle of the sheet of paper that your teacher gave you.
2. Does your organism perform cellular respiration, photosynthesis, or both?

Part I—Matter

Discuss the following questions with your group. Assume your organism is exposed to the Sun.

3. How are carbon atoms entering the organism? (carbon dioxide, food molecules)
4. How are carbon atoms leaving the organism?
5. Use your answers to questions 1 through 4 to make a model of carbon flow through your organism.

On the sheet of paper with your organism's name, show in which ways carbon enters and leaves your organism.

Part II—Energy

Discuss the following questions with your group.

6. What forms of energy can your organism use? (light, energy provided by food)
7. What forms of energy can your organism release? (light, thermal, kinetic)
8. On the sheet of paper with your organism's name, use your answers to questions 6 and 7 to show what forms of energy flow into your organism and what forms flow out of it. Use a different color than the one you used for carbon.

Source: Christine Gleason, Greenhills School. Used with permission.

Assessing Energy Conservation in a Chemistry Context: Photosynthesis and Respiration Reactions

When assessing energy conservation in a chemistry context, we may, for example, look at the chemical processes going on in photosynthesis in greater detail and how these processes link to energy input and output (see Figure 8.6).

Like the Priestley experiment task, this task asks students to create a model of the processes happening in photosynthesis using their understanding of energy as a DCI and CCC. However, this time we do not expect students to link all Five Big Ideas about energy together. Rather, we want students to link the first three Big Ideas—that all energy is fundamentally the same even though it can manifest in different forms (Big Idea 1), that energy can be converted from one form to another (Big Idea 2), and that energy can be transferred between systems and objects (Big Idea 3)—with other scientific ideas, such as with the chemical process by which carbon dioxide and water can be combined to form glucose and oxygen when there is an input of light energy. In middle school, we do not (yet) expect students to provide full details of the chemical processes for photosynthesis or respiration (or burning) in their answer—that is, the exact chemical reaction or quantification of the relationship between the input of energy and the chemical reaction. Instead, we expect students to focus on the general idea of chemical reactions—that is, the rearrangement of the (different) molecules leading to the release of energy (as in cellular respiration and burning) or requiring the input of energy (as in photosynthesis). In their model, we expect students to identify carbon dioxide molecules and water molecules being rearranged into oxygen and sugars under the input of energy from sunlight (i.e., light energy) and the sugars, when oxidized, being rearranged into carbon dioxide and water under the release of kinetic and thermal energy (see Figure 8.6).

Looking Across Disciplinary Contexts

In this section, we have demonstrated three different tasks situated in the disciplinary contexts of physics, biology, and chemistry. Every one of these tasks requires students to develop a model based on their understanding that there is a single quantity called energy, which is conserved as it is transferred and transformed in various ways. That is, students need to connect the Five Big Ideas about energy together as well as to other related scientific ideas. Notice that each task in this section represents a different disciplinary context; across all tasks, students need to demonstrate understanding of the crosscutting nature of energy. No one task can fully assess student understanding, and only by giving students the opportunity to demonstrate their knowledge across a variety of contexts can we get a clearer picture of how well students can consistently apply the Five Big Ideas about energy to make sense of real-world systems.

Figure 8.6. Sample task requiring students to construct simple models of the chemical processes in photosynthesis and cellular respiration (or burning). Notice that sample responses track substances (displayed as names rather than chemical formulas or molecular models) and energy inputs/outputs.

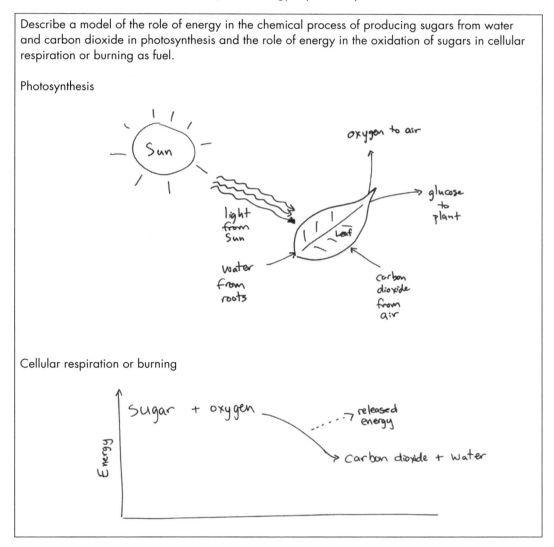

Assessing Students' Understanding of Energy in Photosynthesis

The *Framework* (NRC 2012) describes the process of photosynthesis in this way:

> *Many organisms use the energy from light to make sugars (food) from carbon dioxide from the atmosphere and water through the process of photosynthesis, which also releases oxygen. In most animals and plants, oxygen reacts with carbon containing molecules (sugars) to provide energy and produce waste carbon dioxide. (p. 148)*

The *Framework* also develops a vision of how students are expected to progress toward a full understanding of this learning goal (see Chapter 5). But how can we assess this progression within the course of a year, over a grade band, or across all of K–12? One task that can be used in middle school to assess students' understanding of energy in the context of photosynthesis is the task from Figure 8.6, in which students construct a model to represent the flow of energy and materials. To do this task, students must draw on disciplinary ideas from the life sciences, the CCC of energy and matter flow, and the scientific practice of modeling. But this task is inappropriate for use at the elementary school level. What might an assessment look like to indicate whether elementary students are on track to developing a deep understanding of photosynthesis?

Let us start with revisiting what we expect students to learn in elementary school that is related to this core idea. The *Framework* and the *NGSS* specify that by the end of second grade students should know that plants need water and light to live and grow, and that by the end of fifth grade students should know that animals and plants alike need air and water, animals need to take in food, and plants need light (and minerals). Asking students to draw a model that shows what a plant needs to grow and survive is a useful task for assessing whether early and upper elementary students are adequately progressing in their understanding of photosynthesis. Figure 8.7 shows examples of elementary students' drawings made for this task.

TASK: Plant growth
GRADE BAND: Elementary school
DCIs: Energy (PS3.A, PS3.B, PS3.D); From molecules to organisms: Structures and processes (LS1.C)
CCC: Energy and matter: Flows, cycles, and conservation
PRACTICE: Developing and using models

The first thing to notice about this task is that it allows for models of different levels of complexity. Asking students to draw a model of plant needs allows students to demonstrate different levels of complexity in their understanding. Notice that the model on the left of Figure 8.7 (from an early elementary student) identifies all of the elements that the *Framework* specifies that a student should know by the end of second grade. The model on the right (from an upper elementary student) illustrates what we would expect from a

Figure 8.7. Examples of early (left) and upper (right) elementary students' drawings of what a plant needs to grow and survive

Draw a model that shows what a plant needs to grow and survive.

Source: Drawing on left by Sarah Nordine; drawing on right by Anna Nordine.

student at the end of fifth grade. It shows that plants need light, air, water, and minerals to grow, and where these elements come from.

Often, when assessing elementary students' understanding of energy in the context of photosynthesis, we simply require them to identify the Sun as a source of energy. But this answer omits important aspects of the process by which plants make food. Water, air, and sunlight are all necessary for photosynthesis to occur; the removal of any one of these will stop the process and make it impossible for the plant to make its own food. Further, the *Framework* and the *NGSS* stress the flow of matter over the flow of energy in the elementary grades, and is not until the upper elementary grades that the idea of energy is explicitly included at all. Although the focus in the early grades is on the needs of plants more than the flow of energy, this focus helps set the stage for students to develop a deeper understanding of photosynthesis at the middle and high school level. Elementary students can find evidence that sunlight transfers energy to the plant that is critical for growth, but investigations of the energy transfers and transformations in photosynthesis—and even calling the process photosynthesis—must wait.

In middle school, students can more explicitly investigate the flow of matter and energy in photosynthesis and represent the processes of photosynthesis and respiration in simple models, as the example drawing for the Priestley experiment task in Figure 8.6 shows.

In high school, students are ready to develop a more quantitative understanding of the chemical processes going on in photosynthesis. By the end of 12th grade, students should be able to model how the process of photosynthesis transforms light energy into stored chemical energy. The elodea laboratory task (Appendix 8A) can provide evidence about whether student can gather and analyze data that lead to the development of quantitative models of photosynthesis. In this task, students are required to quantify the energy flow and matter flows in photosynthesis, more specifically the relationship between carbon dioxide intake and the production of oxygen, and they describe the role of energy in this process.

Like the tasks in Figures 8.6 and 8.7, the elodea laboratory focuses on the role of energy and matter in the process of photosynthesis. However, at this age we expect students to show a more sophisticated understanding of the chemical process of photosynthesis and be able to make connections between the macroscopic and microscopic world. Further, we expect that students will connect matter and energy flows in the photosynthesis process. To assess students' performance on the elodea laboratory task, we use a vee diagram (Figure 8.8).

> **TASK:** Elodea laboratory
> **GRADE BAND:** High school
> **DCIs:** From molecules to organisms: Structures and processes (LS1.C); Matter and its interactions (PS1.B); Energy (PS3.A, PS3.D)
> **CCC:** Energy and matter: Flows, cycles, and conservation
> **PRACTICES:** Asking questions, planning and carrying out investigations, analyzing and interpreting data, engaging in argument from evidence

The vee diagram (Gowin 1970, 1981) is a conceptual tool that can help students plan, carry out, and make sense of scientific investigations (see also Liu 2010). The essential components of a vee diagram are (a) focal questions, (b) events and objects, (c) theories and principles, (d) concepts, (e) records and transformation, and (f) knowledge claims.

In the elodea laboratory, the vee diagram aids students in conceptualizing the mechanisms involved in photosynthesis and the role that energy plays in the process. The students complete the vee diagram in three steps:

1. Students make their prior knowledge about photosynthesis explicit by completing the "Conceptual" side (left side) of the Vee diagram. They complete the "What do I know?" and "How are the ideas connected?" sections of the diagram.

2. Students perform the elodea laboratory task, in which they investigate the effect of light intensity on the rate of photosynthesis by measuring the amount of oxygen produced (see Appendix 8A). Students are given copies of the elodea laboratory and asked to read it over and complete the "Events" section of the vee diagram. Students then work through the laboratory exercise.

3. Students complete the "Methodological" side (right side) of the vee diagram by recording their data collection and claims. To complete this side of the diagram, students consider both the conceptual side and the observations and data they recorded in their laboratory worksheet.

Figure 8.8. Vee diagram for elodea laboratory

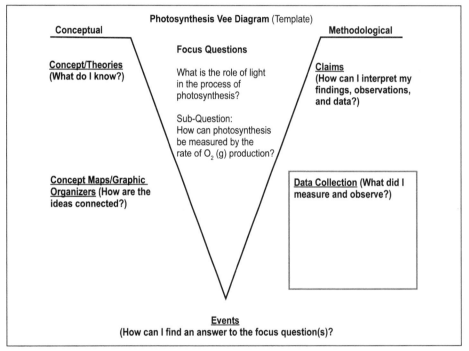

Source: Adapted by Erica Smith, Cuba-Rushford High School, New York, from Doran et al. 2002.

The vee diagram is intended to support students in making connections between the laboratory experiences and their understanding of photosynthesis. In particular, students use the vee diagram to represent their understanding of photosynthesis with regard to light as an energy source, the transfer of energy into the system, and the production of oxygen gas. Although the vee diagram is primarily a support for student thinking, it also serves as a powerful assessment opportunity. The scoring rubric in Table 8.1 (p. 158) gives an example of how the vee diagram may be analyzed.

The elodea laboratory is surely a complex task, but it is accessible for high school students and tests for a deep understanding of photosynthesis. Solving the task involves not only a deep understanding of photosynthesis but also competency with different scientific practices, among them Asking Questions, Planning and Carrying Out Investigations,

Table 8.1.

SCORING RUBRIC FOR THE ELODEA LABORATORY VEE DIAGRAM

Object/event	Points
No objects of events are identified.	0
The major event or objects are identified and are consistent with the focus question, or an event and objects are identified but are inconsistent with the focus question.	1
The major event and the accompanying objects are identified and are consistent with the focus question.	2
Same as above, but also suggests what observations and data will be collected.	3

Principles and concepts	Points
No information is presented on the conceptual side.	0
A few concepts are identified or a principle written is a knowledge claim sought in the investigation.	1
Concepts and at least one type of principle (conceptual or methodological) or concept and relevant theory are identified.	2
Concepts, at least two types of principles, and relevant theories are identified.	3

Concept map	Points
No concept map is drawn.	0
Concept map is drawn and includes a few major concepts that are relevant to the focus question, but it is not hierarchical and does not include linking words between concepts.	1
Concept map is drawn and includes a few major concepts that are relevant to the focus question, is in a hierarchical format, and includes linking words between concepts.	2
Same as above, but also includes cross-links between concepts with appropriate linking words.	3

Records/transformation (data collection)	Points
No records or transformations are identified.	0
Records are identified but are inconsistent with the focus question of the major event.	1
Either records or transformations are identified but not both.	2
Records are identified for the major event; transformations are consistent with both the focus question and the abilities and grade level of the student.	3

Knowledge claims	Points
No conclusions are made.	0
Conclusions are made but are wrong.	1
Conclusions are partially correct.	2
Conclusions are made based on data and are correct.	3

Source: Adapted from Doran et al. 2002 and Liu 2010.

Analyzing and Interpreting Data, and even Engaging in Argument From Evidence. As such, the task assesses many of the most important aspects of the vision of science learning described in the *Framework*.

It is important to note that students in the elementary and middle school grades are simply not ready to make sense of the elodea laboratory; the tasks in Figures 8.6 and 8.7 represent intermediate steps that can indicate whether students are on track to develop the kind of three-dimensional understanding that is required for success on something like the elodea laboratory task when they get to high school. In elementary school, students are expected to model the energy and matter flow into a plant; in middle school, students are expected to describe the flow of matter and energy at the microscopic level; and in high school, students are ready to conduct a more quantitative analysis of this flow. By focusing on tasks that are appropriate for each grade level while also focusing consistently on developing the Five Big Ideas of energy over time for the same ideas, students can be well positioned to develop understandings that are both accurate and applicable across a wide variety of contexts.

Assessing Students' Understanding of Natural Resources

So far in this chapter, we have discussed tasks intended for assessing students' understanding of energy as a CCC and their ability to apply this understanding involving various scientific practices across different disciplinary contexts and grade bands. To assess whether students are developing a crosscutting understanding of energy, it is important both to look for evidence of whether students invoke the Five Big Ideas of energy as they complete disciplinary tasks and to provide students with opportunities to consistently use their understanding of energy across a wide range of contexts. Obviously, no single task will give us a complete picture of student understanding, but some contexts are better positioned than others to help students demonstrate their understanding of energy across disciplines. Human use of natural energy resources can provide such a cross-disciplinary context.

The *Framework* (NRC 2012) specifies that students should come to know these ideas about natural resources:

> *All materials, energy, and fuels that humans use are derived from natural sources, and their use affects the environment in multiple ways. Some resources are renewable over time, and others are not. (p. 192)*

The task shown in Figure 8.9 (p. 160) is an example of an assessment activity that can elicit students' ideas about energy in the context of human use of natural energy resources. The task elicits students' understanding of the Big Ideas about energy and provides them with opportunities to demonstrate their understanding of the connections among these ideas as they work through the task.

Figure 8.9. A natural energy resources task focusing on using soy-diesel to power buses (sample answers in italics below)

The following figure shows a bus that uses soy-diesel instead of gasoline. Soy-diesel is made from soybean oil, which is extracted from the beans of the soybean plant.

1. Draw a model of the energy transformations that occur when the soy-diesel is burned in the bus's engine, causing the wheels of the bus to turn.

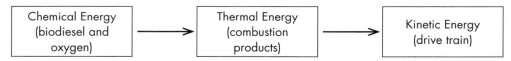

2. Provide an explanation of why energy is released when soy-diesel is burned.

When soy-diesel is burned with oxygen, energy is required to break the bonds in the soy-diesel and in the oxygen. When new bonds are formed between the atomic constituents of soy-diesel and the oxygen, energy is released. The amount of energy released in the formation of the new bonds is greater than the amount of energy required to break the original bonds, so, altogether, energy is released in the process.

3. Design an experiment which would allow you to estimate the amount of energy that can be obtained by burning one soybean.

Burn a soybean to heat an ounce of water; measure the temperature of the water before and after burning the soybean.

4. Draw a model that shows where the energy required to create the bonds in the soybean molecules came from. Be sure that your model represents both matter and energy changes.

Models should show that inputs of carbon dioxide, water, minerals, and sunlight are required for plant growth. Model outputs should include glucose and oxygen, and may represent more complex macromolecules in the soybean plants. Models should identify sunlight as the source of energy for creating the bonds in the soybean molecules.

Figure 8.9 (*continued*)

5. Use the diagram to explain why reducing the weight of the bus can help save energy!

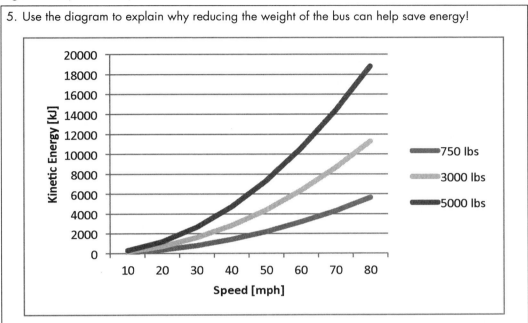

The diagram shows that to accelerate an object to a given speed, we need to transfer energy to it. The speed depends on the energy transferred to the object and the object's mass. According to the diagram, the greater the mass of the object, the more energy is needed to accelerate the object to the same speed. Thus, if the bus is constructed from lighter materials (aluminum or carbon), less energy is needed to accelerate it to the same speed.

Image source: User: Adrignola, https://upload.wikimedia.org/wikipedia/commons/5/5c/Soybeanbus.jpg

You will notice by reading through the task in Figure 8.9 that individual items also require students to engage in a set of scientific practices. For example, item 1 asks students to model the energy transfer and transformation processes in our real-world situation—the bus running on soy-diesel. Notice that this item is very similar to the Rube Goldberg machine task shown in Figures 8.3 and 8.4. In item 2, students construct an explanation for how energy is released when soy-diesel is burned. Here, we expect students to go beyond the simple recalling of a single transformation process of chemical energy of the fuel (and the

> **TASK:** Soy-diesel bus
> **GRADE BAND:** High school
> **DCIs:** From molecules to organisms: Structures and processes (LS1.B, LS1.C); Matter and its interactions (PS1.A, PS1.B); Energy (PS3.A, PS3.B, PS3.D)
> **CCC:** Energy and matter: Flows, cycles, and conservation
> **PRACTICES:** Developing and using models, constructing explanations, planning and carrying out investigations

oxygen) into kinetic energy and exhibit an understanding of physical systems by identifying the connection between fuel and oxygen and how the rearrangement of molecular bonds connects to the car moving. In item 3, students are required to plan an investigation. Items 4 and 5 require students to develop a model and construct an explanation from data. Each component of this assessment task engages students in a scientific practice as they use their understanding of energy to make sense of the process of using soy-diesel to power a bus.

Every item in the soy-diesel bus task requires students to connect at least several, if not all, of the Five Big Ideas about energy (see the sample answers provided in Figure 8.9). Item 4, for example, prompts students to identify the energy source for the process of photosynthesis and to model critical matter and energy changes during photosynthesis and food production in plants.

To correctly respond to item 4, students need to apply the principle of energy conservation (Big Idea 4) to correctly identify how the energy flows (transfers, Big Idea 3) are connected to the processes in the plant. Likewise, item 5 prompts students to connect the idea of kinetic energy (Big Idea 1) to the idea of energy transformation and transfer (Big Ideas 2 and 3) and to use the ideas of conservation and dissipation (Big Ideas 4 and 5) to explain how reducing the weight of a bus can help in saving energy. Item 5 also gives students an opportunity to demonstrate an understanding of the difference between the scientific principle of energy conservation and the everyday idea of conserving (i.e., saving) energy resources.

The soy-diesel bus task elicits students' understanding of the Five Big Ideas about energy and their ability to connect these ideas together to solve authentic problems. Throughout the task, students also use different scientific practices. Finally, the task is situated in an authentic real-world context that integrates energy ideas from multiple disciplines. As such, the soy-diesel bus task provides a good example of how students' three-dimensional learning about energy can be assessed. Assessment contexts that focus on energy resources are a natural way to connect energy ideas across multiple disciplines because these contexts can easily include prompts related to photosynthetic processes in plants (biology) and using energy stored in food and fuels (chemistry) to power machines (physics).

The soy-diesel bus task is appropriate for use with high school students, but the soy-diesel bus context can be used with elementary and middle school students to emphasize key ideas at each grade band. In elementary school, students can be prompted to (1) think about where the energy for moving a soy-diesel bus came from in the first place (light from the Sun), (2) trace the series of events necessary (photosynthesis, plant growth, harvesting soybeans for fuel, burning fuel) to provide energy for the bus to move, and (3) connect these events together using the idea of energy transformations and transfers. In middle school,

> **TASK:** Soy-diesel bus
> **GRADE BAND:** Elementary school
> **DCIs:** From molecules to organisms: Structures and processes (LS1.B, LS1.C); Energy (PS3.A, PS3.B, PS3.D)
> **CCC:** Energy and matter: Flows, cycles, and conservation
> **PRACTICE:** Developing and using models

students can develop a more sophisticated representation of the energy forms
and transfers involved in powering a
bus with soy-diesel; items 1, 3, and 5
might be altered very little for middle
school students, but the criteria of a
complete answer would be modified to
emphasize the ideas that middle school
students should know according to the Framework.

> **TASK:** Soy-diesel bus
> **GRADE BAND:** Middle school
> **DCIs:** From molecules to organisms: Structures and processes (LS1.B, LS1.C); Matter and its interactions (PS1.A, PS1.B); Energy (PS3.A, PS3.B, PS3.D)
> **CCC:** Energy and matter: Flows, cycles, and conservation
> **PRACTICES:** Developing and using models, constructing explanations, planning and carrying out investigations

The use of natural energy resources provides an assessment context that is comprehensible to a wide range of students, easily adaptable for use at different grade bands, and
inclusive of all Five Big Ideas about energy. As such, this context provides an opportunity to
track students' growth in energy understanding over time and across disciplines. Whether
you use such tasks as a formative or summative assessment within the course of a unit or
a school year or across multiple years, tasks that focus on natural resources are a powerful
opportunity to gauge student understanding within a meaningful real-world context.

Assessing Three-Dimensional Learning About Energy

The tasks presented in this chapter combine a variety of scientific practices and core ideas
within both disciplinary and cross-disciplinary contexts. You may notice that the tasks do
not address specific *NGSS* performance expectations. This is intentional. We did this to
emphasize that students' learning about energy and assessment of student learning is not
only about the (relatively few) performance expectations that can be found in the *NGSS*.
The performance expectations in the *NGSS* define what students are expected to be able
to do at the end of the respective grades or grade bands. They represent the knowledge
and skills on which students will be assessed. This, however, does not mean that students
should only be able to meet these performance expectations or should be trained to specifically meet these performance expectations. Instead, the performance expectations in the
NGSS are intended to span the space of the competence in science that students are expected
to have developed at the end of selected grades or grade bands. The examples presented
in this chapter are intended to illustrate the variety of tasks that can be used to foster and
assess students' three-dimensional learning on their way to meet these expectations.

No assessment task is perfect, and that goes for the examples in this chapter as well.
Even if an assessment works very well in one setting, it may be a flop in another. For example, the soy-diesel task may interest students in a location where bus riding is common and
alternative energy resources are a frequent topic of discussion, whereas students living in
areas where bus riding is rare may find the task less engaging and comprehensible. When
choosing assessment contexts, it is important to tap into situations that have relevance for

your students and to ensure that assessment contexts are reasonably comparable to the instructional contexts that you have chosen to include in your classroom. An incomprehensible task context will provide you with little accurate information about students' understanding of energy. In addition to picking the right context, assessment tasks should be constructed to ensure that students are truly activating their own ideas and that students are in a position to demonstrate a three-dimensional understanding of energy.

As you design your own assessment tasks for use with your students, the checklist in Box 8.1 may be helpful. Although there is no perfect item and no assessment will fully exemplify every item in the checklist, good assessments are strong across a variety of the items listed in Box 8.1.

Box 8.1. Checklist for designing assessment items for three-dimensional learning

---

Does my assessment ...

1. have an authentic, real-world, and multidisciplinary context?

2. require linking multiple ideas about energy (and other core ideas)?

3. require the application of multiple practices?

4. combine the application of scientific practices and multiple ideas about energy (and other core ideas) in the context of authentic, real-world, and multidisciplinary contexts?

5. incorporate a developmental perspective?

6. provide sufficient information to plan the next steps of learning?

7. include multiple response formats?

A critical recommendation in the *Framework* and the *NGSS* is that instruction should emphasize science and engineering practices, DCIs, and CCC. If we make this change in instruction but fail to include a similar emphasis in both formative and summative assessments, then students will soon get the message that what's really important is the reproduction of isolated facts and simple procedures—a message that has been too common in too many science classrooms. High-quality assessments are just as critical as instruction in driving meaningful three-dimensional learning.

Summary

When students leave school, we hope that they will be able to make use of what they have learned to make sense of real-world situations and make well-informed decisions. Although students learn about many scientific concepts in school that they may find difficult to apply in real-world settings, the energy concept is ubiquitous in both scientific and nonscientific settings. Some of the most important decisions that students will make in their lives are energy related: What car to buy, what to eat, and the natural resources we should use are all decisions that are fundamentally connected to the energy concept. Yet, school science instruction often fails to present and assess the energy concept within meaningful real-world contexts, and existing assessments often emphasize recall of isolated facts above application and connection of multiple ideas about energy to make sense of meaningful scenarios.

The Five Big Ideas about energy help teach energy in a way that is consistent across disciplines, and they provide a lens for assessment as well. Coupled with the principle of three-dimensional learning (integrating scientific practices, DCIs, and CCCs), the Five Big Ideas of energy provide a framework for assessing students' ability to apply energy ideas consistently across science disciplines and within real-world contexts.

It takes years of effortful practice for students to gain competency with scientific concepts and practices, and this is especially true for the energy concept. The *Framework* and the *NGSS* provide a roadmap for how students can develop their ideas about energy over time and apply them in increasingly sophisticated ways, and these documents guide classroom assessment just as much as they inform classroom instruction. Because energy is a disciplinary core idea, a crosscutting concept, and critical to everyday life, it provides an excellent opportunity to implement science assessments that are three-dimensional in nature and situated within meaningful contexts.

References

Aronson, E., and S. Patnoe. 1997. *The jigsaw classroom: Building cooperation in the classroom.* 2nd ed. New York: Longman.

Doran, R., F. Chan, P. Tamir, and C. Lenhardt. 2002. *Science educator's guide to laboratory assessment.* Arlington, VA: NSTA Press.

Gowin, D. B. 1970. The structure of knowledge. *Educational Theory* 20 (4): 319–328.

Gowin, D. B. 1981. *Educating.* Ithaca, NY: Cornell University Press.

Liu, X. 2010. *Essentials of science classroom assessment.* Los Angeles: SAGE.

National Research Council (NRC). 2012. *A framework for K–12 science education: Practices, crosscutting concepts, and core ideas.* Washington, DC: National Academies Press.

National Research Council (NRC). 2014. *Developing assessments for the Next Generation Science Standards.* Washington, DC: National Academies Press.

Neumann, K., T. Viering, W. J. Boone, and H.E. Fischer. 2013. Towards a learning progression of energy. *Journal of Research in Science Teaching* 50 (2): 162–188. *http://doi.org/10.1002/tea.21061.*

NGSS Lead States. 2013. *Next Generation Science Standards: For states, by states.* Washington, DC: National Academies Press. *www.nextgenscience.org/next-generation-science-standards.*

APPENDIX 8A.
ELODEA LABORATORY

Photosynthetic organisms (cyanobacteria) first evolved about 3.5 billion years ago; the oldest known fossils on Earth are of cyanobacteria. Cyanobacteria live in water, can manufacture their own food, and are one of the most important groups of bacteria on Earth. Cyanobacteria have been important in shaping the course of evolution and ecological change throughout Earth's history. Through photosynthesis, cyanobacteria take in atmospheric carbon dioxide, water, and sunlight and convert it into sugar (glucose), releasing oxygen.

Photosynthesis reduced the amount of carbon dioxide in the atmosphere, as oxygen was continually being released. For roughly a billion years, oxygen released by cyanobacteria did not build up in the atmosphere. It was used up by two sources: oceans and rocks. Oxygen dissolved in the oceans and oxidized (rusted) the exposed iron and other minerals, as seen in banded rock formations around the world. Approximately 2 billion years ago, the reservoirs of oxidizable rock became saturated and, thus, allowed the buildup of oxygen in the air.

As oxygen moved into the early atmosphere, ultraviolet radiation from the Sun split the oxygen molecules (O_2), which then recombined, producing the Earth's ozone layer (O_3). This, in turn, reduced the amount of incoming ultraviolet radiation striking the Earth. The impact for life on Earth was enormous. With reduced amounts of ultraviolet radiation, organisms moved to shallow water and, ultimately, onto the land.

Safety Notes
1. Wear sanitized indirectly vented chemical-splash goggles during pre-lab setup, the activity, and post-lab cleanup.
2. Review the safety data sheet for baking soda with students.
3. Use only a GFI or GFCI power source for the lamp.
4. Caution students not to touch the lamp when it is hot—burn hazard!
5. Review with students the technique for using a razor blade. Remind them it is a sharp hazard and can seriously cut skin!
6. Wipe up any water spilled on the floor—slip-and-fall hazard!
7. Wash hands with soap and water after completing the lab.

Materials

- Elodea stem (up to 10 cm, or 4 inches)
- Distilled water
- Baking soda
- Beaker
- Lamp (60–100 watt)
- Metric ruler
- Large test tube
- Single-edged razor blade
- Timer or clock

Source: Erica Smith, Cuba-Rushford High School, New York. Used with permission.

Appendix 8A (*continued*)

Research Question
This lab will allow you to observe the release of oxygen gas by a freshwater plant, elodea. You will manipulate the conditions for photosynthesis by changing amounts of available light. Can changes in light influence the rate (speed) of photosynthesis?

Purpose
The purpose of the lab activity is to observe photosynthesis in action using the freshwater plant elodea. Photosynthesis converts carbon dioxide, water, and energy (in the form of light) to sugar, water, and oxygen. The chemical equation looks like this:

$$6 \; CO_2 + 12 \; H_2O \rightarrow C_6H_{12}O_6 + 6 \; H_2O + 6 \; O_2$$

Procedure
Station Setup

1. Obtain an elodea stem. Carefully cut away several leaves from the cut end of the stem. Approximately 5 mm (1.5 inches) from the cut end, slice the stem at an angle. Lightly crush this end with your fingers.
2. Place the stem in a test tube, partially filled with water, cut end up. Fill the test tube to near the top with water.
3. Place the test tube in the beaker.

Running the Activity
1. Place a lamp 5 cm (2 inches) from the test tube.
2. Wait for the first signs of bubbles to begin the activity. If no bubbles, cut and gently crush the stem again.
3. Once bubbles start to appear, record the number of bubbles that appear during a three-minute interval. Repeat recording bubbles for three minutes at least one more time. Average your numbers.
4. Move the lamp back 5 cm, 10 cm, 15 cm, and 25 cm (2 inches, 4 inches, 6 inches, and 10 inches) or as determined by your teacher. Wait one minute and record the number of bubbles that appear during a three-minute interval for each light adjustment. Record your numbers on the board under each distance.
5. Fill in the class data and calculate the average number of bubbles for each distance.
6. Add a pinch of baking soda to the test tube. Move the lamp back to the 5 cm position. Wait one minute and record the number of bubbles that appear during a three-minute interval. Repeat recording bubbles for an additional three minutes. Average your numbers.
7. Move the lamp back 5 cm, 10 cm, 15 cm, and 25 cm. Wait one minute and record the number of bubbles that appear during a three-minute interval at each location. Record your numbers on the board under each distance.
8. Fill in the class data and calculate the average number of bubbles for each distance.

Appendix 8A (*continued*)

Data Collection

Distance from light source	# of bubbles in H_2O	# of bubbles in baking soda
5 cm (Your Number)		
5 cm (Class Data)		
Average at 5 cm		
10 cm (Your Number)		
10 cm (Class Data)		
Average at 10 cm		
15 cm (Your Number)		
15 cm (Class Data)		
Average at 15 cm		
25 cm (Your Number)		
25 cm (Class Data)		
Average at 25 cm (10 in)		

Observation and Analysis of Results

1. What did you observe? Compare the water observation to the baking soda observation.

   Interpret and analyze your results by answering the following questions:

2. On the graph, use your data to construct a bar graph of your control data (with water) and experimental data (with the baking soda). Be sure to label the graph and give the graph an appropriate title.

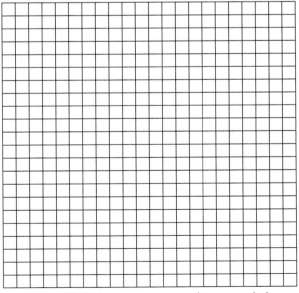

y = # of bubbles x = distance to light

Appendix 8A (*continued*)

1. Why did you move the lamp away from the elodea stem?
2. Why did you add baking soda?
3. What is the independent variable for your group?
4. What change did you expect in the control group?
5. What change did you expect in the experimental group?

List the biotic and abiotic factors present.

SECTION 3

SUPPORTING TEACHERS IN EMPHASIZING ENERGY AS A CROSSCUTTING CONCEPT

CHAPTER 9

PROFESSIONAL DEVELOPMENT FOR TEACHING ENERGY

ROBERT CHEN AND ARTHUR EISENKRAFT

How Do We Support Teachers in Teaching Energy as a Crosscutting Concept?

Imperatives

A Framework for K–12 Science Education (National Research Council [NRC] 2012) and the *Next Generation Science Standards* (*NGSS*; NGSS Lead States 2013) call for new ways to teach students about energy that will require fundamental shifts in how teachers present energy ideas to students. To support teachers in making these shifts, we must consider the entire K–12 educational system because students' experience with energy as a disciplinary core idea (DCI) and as a crosscutting concept (CCC) continues throughout their K–12 educational experience. That said, it is not the responsibility of every elementary, middle, and high school teacher to know and teach everything there is to know about energy, but it is important for every teacher to know how his or her teaching of energy fits into the overall learning progression for students. Therefore, it is critical that professional development (PD) for teachers occur at all grade bands and that professional developers and teacher educators consider connections between grade bands and disciplines as they design energy-related instructional supports for teachers.

Energy plays a central role in both the *Framework* and the *NGSS*. As a DCI, energy must be considered in a variety of contexts. In physics, energy is connected to forces and motion, the correspondence between macroscopic and microscopic perspectives, and its relation to fields. In chemistry, energy in bonds governs the total release or absorption of energy in chemical reactions. In biology, energy flows through ecosystems but is converted from "high-quality" sources of energy (such as food and fuel) to less useful forms (such as infrared radiation leaving the Earth). In Earth science, energy is a natural resource that has guided the development of human society; most of that energy comes from the Sun. In addition to its central role in each discipline, or rather *because* of it, energy is also a CCC that allows scientists to use energy in consistent ways across disciplinary boundaries.

The Five Big Ideas of energy (see Chapters 1 and 2) are useful for thinking about the role of energy in a wide variety of disciplinary and interdisciplinary contexts. In addition to its central role in the sciences, energy is a major social issue in that the global society will need three times as much energy in 2100 as it uses today (Intergovernmental Panel on Climate Change 2000). Our present reliance on fossil fuels is not sustainable because these fuels are not renewed at a rate sufficient to support such an increased demand. Further, these fuels emit carbon dioxide gas, which contributes to the warming of Earth's atmosphere over time. Because the consequences of our global energy resource supply and use will affect fundamental aspects of our lives, there has been a campaign from the U.S. Department of Energy to strive for everyone to become energy literate (see Chapter 7). Thus, the benefits of teaching energy will extend far beyond the classroom walls. Designing and implementing energy instruction that spans the disciplines and also extends into students' lives outside of school is no small task, and teachers need access to PD that clarifies the enhanced role that energy plays in the *NGSS* and provides opportunities for working with colleagues to enhance existing energy instruction.

Constraints

Every PD experience is limited by time and available resources, and energy PD is no exception. There is a lot to learn: energy as a DCI, energy as a CCC, how energy connects to key scientific and engineering practices, available instructional technologies, and pedagogical content knowledge (PCK) for teaching energy, to name a few. Often, content-based PD is constrained to disciplinary experiences focused solely on the physical sciences, the life sciences, or the Earth sciences, or even computer sciences or engineering. However, few of these disciplinary experiences support the development of teaching with CCCs in mind. For example, biology students learn about photosynthesis and respiration and are usually quite capable of responding to questions regarding these DCIs. Yet, when students are asked to discuss energy as it relates to photosynthesis and respiration, as they were on the first administration of the revised Advanced Placement (AP) Biology exam in 2013, many students stumble (Domenic Castignetti, Chief Reader AP Biology, personal communication). Likewise, students may accept the idea of energy conservation in the disciplinary context of the physical sciences but contradict this law when making sense of biological systems (Chabalengula, Sanders, and Mumba 2012). How can PD address all of the relevant DCIs, scientific practices, and PCK, and now add CCCs as well as engineering practices? It can't—there simply is not enough time to do it all!

The last thing that teachers need is yet another thing to be sure that they must emphasize in their instruction. But, if done right, energy PD can clarify—rather than complicate—teachers' existing science instruction. Just as energy is a unifying principle in the sciences, it can be a powerful venue to illustrate how DCIs, CCCs, and practices can be addressed in concert with one another. Although its crosscutting nature makes energy a powerful unifying idea, it is not without its complications. Unlike CCCs such as Patterns and Cause and Effect, energy is both a CCC and a DCI, and this dual role may be confusing. Further, the energy concept itself is nuanced, abstract, and heavily influenced by our everyday experiences using the word *energy* outside of scientific contexts.

As discussed in Chapter 2, students come to the classroom using the word *energy* and related words (e.g., *work* and *power*) in their everyday language, and the everyday usage of these words may not correspond with scientific ideas. Most teachers do the same thing, and if they do not have a strong science background, they may never have considered the dichotomy between the everyday and scientific meanings of energy-related words. While the prevalence of energy in our everyday lives is useful for making connections beyond the classroom, it also introduces challenges associated with using the same words in both scientific and nonscientific contexts. To teach words that are used differently in the vernacular, teachers need to become familiar with the scientific use of the terms. They must also become aware of common student preconceptions and develop the PCK to be able to leverage students' intuitive ideas to help them develop more scientifically oriented ways of looking at phenomena. If teachers develop strategies for helping students make connections among the various ways of using the energy concept, then their students will be in a much better position to use energy to reason about an amazing diversity of problems and relevant scenarios.

In addition to the challenges associated with understanding and teaching about energy as both a DCI and a CCC, it may be unclear to many teachers how participating in PD emphasizing energy as a CCC will help them improve their students' scores on high-stakes accountability tests. Like teachers, school administrators are under pressure to perform on accountability measures and may find it difficult to support PD initiatives that do not have a clear and direct link to the knowledge and skills on which students will be assessed. Although the idea of assessing energy as a core idea is familiar, the role of energy as a CCC on high-stakes assessments is new.

There are many challenges associated with designing, motivating, and delivering energy PD, and these challenges align with broader issues related to teaching in new ways that correspond to the vision of the *Framework* and the *NGSS*. Energy PD presents some wonderful opportunities for introducing new and important ideas for the teaching and learning of science.

Opportunities

Because one goal of education is to prepare an educated citizenry to participate fully in our democratic society, it is worthwhile to focus on major issues facing our nation and the world. Energy resources, the relation between energy needs and our economy, and the environmental impact of energy-related decisions are important issues today and will almost certainly remain so in the world that today's students will inherit. The importance of energy provides an opportunity to attract teachers to PD that will give them the tools to engage students in the study of energy in their respective science classes. Such PD may help teachers ensure that their students understand the science of energy before being confronted with the politics of energy.

The critical importance of energy in a variety of scientific and nonscientific settings provides a strong rationale for returning to energy throughout a school year and from one school year to the next. As students move from kindergarten to high school, they can be reminded of what they have previously learned about energy and build their new knowledge on this foundation. Students in early grades will learn that light, sound, and motion are related to a scientific concept called energy. They will also learn that energy derived from oil or natural gas or wood can be used to heat our houses. In later grades, they will be able to quantitatively compare the amount of energy in food, oil, and the sunlight we receive and use ideas related to transformation, transfer, conservation, and dissipation to evaluate how the availability of various energy sources affect living and nonliving systems.

Perhaps due to its ubiquity in scientific and everyday settings, energy is already a common topic in science instruction. An analysis of 306 curriculum lessons in the Boston Public Schools science curriculum (grades 1–12) revealed that 105 of these lessons involved energy and that teachers could identify 162 connections among them (Chen et al. 2014). Similarly, the Boston Science Partnership offered 11 graduate-level science content courses, but many of the teachers that took the Energy I course (described later in this chapter) considered it the most important and most effective of any of these normally disciplinary courses.

Energy offers a flexible approach to connecting the DCIs among the various grade bands. Teachers can choose their favorite examples or experiments, they can emphasize the areas of science where they feel most comfortable, and they can use students' individual and collective experiences to support teaching of the Five Big Ideas about energy. In addition, energy provides a strong central concept to engage teachers of all grades to discuss what students have learned, what they are learning, and what they will learn. The vertical articulation of scientific concepts and the connections across the sciences can be bridged through energy. Students' understanding of energy evolves during their K–12 classroom experience, and energy PD that includes elementary, middle, and high school teachers enhances the ability of teachers to address the vertical nature of student learning. Finally, energy is naturally interdisciplinary, allowing complex, real problems in society, the field, or the lab to be addressed in a holistic way. Energy offers an authentic approach to analyzing complex systems. As such, energy PD allows teachers to better address the questions about the real world that come from their students. Overall, the opportunities for energy PD to be efficient and effective for teachers of all disciplines and grade bands far outweigh the constraints.

Features of Effective Energy Professional Development

There are four common features of effective energy PD:

1. It is based on real-world phenomena.

2. It connects grade bands and disciplines.

3. It emphasizes the use of energy as an analytical tool.

4. It emphasizes three-dimensional learning.

Like high-quality classroom instruction, effective energy PD is grounded in explorations of meaningful real-world phenomena. That is, energy PD should focus on natural and designed systems that teachers and students experience both in and out of school. By connecting classroom learning with real-life examples, energy PD provides engaging, authentic learning about how the energy concept helps students understand the world around us. Effective energy PD helps teachers connect school and everyday experiences with energy.

In addition to connecting in-school and out-of-school experiences, effective energy PD helps bridge different disciplines and grade bands. To make such connections, it is useful to involve PD designers and facilitators with backgrounds in different scientific disciplines and experience teaching a range of grade levels. Although such a multidisciplinary collaboration adds another layer of complexity to planning and offering energy PD, it also offers the opportunity for PD facilitators to learn from one another, increase his or her capacity, and ensure that the district curriculum is both coherent and vertically aligned.

Because energy is so widely applicable across both content and grade levels, it is important for teachers to have opportunities within energy PD to use energy as an analytical tool to understand a variety of systems. The more practice that teachers have in using energy to analyze systems in biology, chemistry, physics, Earth science, engineering, and everyday life, the better they will become at consistently applying the Five Big Ideas about energy. As their comfort level grows, so, too, will their ability to pick examples that might best engage their students and to develop strategies for supporting student learning about energy.

Finally, effective energy PD integrates the three-dimensional learning of the DCIs, CCCs, and practices identified by the *NGSS*. As both a DCI and a CCC, energy is a sensible starting place for helping teachers unpack what it means to teach science with practices, DCIs, and CCCs in mind. By integrating all three strands of the *NGSS*, energy PD may serve as an exemplar for understanding and implementing the *NGSS* in schools.

Examples of Effective Energy Professional Development

Boston Energy in Science Teaching: Energy I

PD was a key feature of the Boston Energy in Science Teaching (BEST) project, a National Science Foundation–funded Phase II Math and Science Partnership. BEST was a follow-up to our Boston Science Partnership NSF grant where the Energy I course was first introduced. With a goal of teaching students how to connect the sciences from an energy perspective, BEST leadership offered a variety of energy-related programs. Two graduate courses for inservice teachers, Energy I: Integrating the Sciences Through Energy and Energy II: Using Energy in the Classroom, were fundamental to increasing teacher content knowledge and PCK.

Energy I was designed to increase teacher energy content knowledge across all of the disciplines—physics, chemistry, biology, and Earth science—and to seek connections among the disciplines and connections to real-world phenomena in everyday life. As such, the course was team-taught. The original team was made up of three university professors—a chemical oceanographer, a physicist, a biologist—and a middle school science teacher from the Boston Public Schools. The course has been taught a total of 10 times at the time of writing, with different teams of professors at two institutions and in different formats (one evening per week throughout a semester, two-week summer course); the model has proved to be robust and independent of the specific professors. Several key features make the course successful: team teaching, differentiated learning, phenomena-based learning, an Energy Wall, and a focus on systems.

Team Teaching to Cross Disciplines

Having all the professors and the middle school teacher present throughout each session was critical to smooth integration among the disciplines and maintaining a K–12 classroom focus. Although professors (and teachers) had a variety of discipline-based backgrounds, constant questioning and interjections by instructors and students with energy connections moved the learning from a traditional disciplinary perspective to a CCC perspective. For example, when the physicist presented heat transfer by conduction, he presented a simple diagram and equation to quantitatively describe the process. The chemical oceanographer noted that heat transfer from a warm surface layer to the colder deep ocean was enhanced by salt fingers—small eddy currents driven by density differences—but that the process could still use the same conduction equation, although certain assumptions had to be made. The biologist suggested that conduction never really is simple in biological systems because fur in animals sets up temperature gradients and circulation can regulate initial temperature differences in response to behavior. In this way, teachers not only learned about the concept of heat transfer but also learned how the concept can be applied to multiple situations in very different contexts.

Throughout the course, there was consistent and spirited dialogue among the professors and teachers in the course. It was common for one professor to challenge another or find fault in the simplicity of the model being presented; on the other hand, when a professor focused on the complexity of a system, he or she might be accused of placing too much emphasis on inconsequential amounts of energy. Although this lack of a consensus view was not planned (but also was not an accident), it was a strength of the course because it allowed the teachers taking the course the opportunity to observe highly competent scientists disagreeing, reconsidering, and reformulating. Some teachers reported that this shifted their view of science as a body of facts to one of theories and models. In addition, the limitations of knowledge of the professor instructors led to the healthier perspective that we are all learners. Both of these lessons are important for teachers to communicate to their students. Without different disciplinary perspectives being presented in such close proximity, much of the ability to apply energy concepts across disciplines would be lost.

Differentiated Learning to Connect All Grade Bands and Disciplines

Participants in the Energy I course were inservice teachers working in the Boston area. They were spread across all disciplines and from all grades (K–12) and included AP science teachers. This diversity was critical to using energy to cross grade bands and disciplines, but it could have created problems for teaching the course. By focusing the learning on both the energy concept and the application of that concept in a variety of classrooms and everyday life situations, all the teachers taking the course were able to participate. In addition, the large disparity in teachers' comfort with mathematics was addressed by solving problems in self-differentiated groups (beginning, intermediate, advanced), using a set of 20 problems that advanced from simple to difficult. All participants were, therefore, engaged in all activities, problem-solving sessions, and discussions, and all were pushed to learn at their own levels.

Phenomena-Based Learning

Many of the activities and discussions were centered on real-world phenomena that were familiar to all teachers or that we could share as a common experience in the class. The course started with dropping a tennis ball to explore conversion of gravitational potential energy to kinetic energy to elastic potential energy, with the realization that some of the energy was dissipated so that the ball would not bounce all the way back to its original height. By exploring many different bouncing balls and losses of energy, teachers shared common experiences that could be used to ground learning for any energy conversion in any system. For example, teachers created a battery-and-bulb electrical circuit. They initially interpreted this as chemical energy in the battery lighting the bulb. Using the bouncing ball as a model, they were able to refine this explanation to something that includes the loss of potential energy in the chemicals of the battery providing for an increase in the kinetic energy of the molecules in the bulb's filament. The increase in kinetic energy was observed as an increase in temperature, which results in the bulb emitting light. Not all of the chemical energy was transferred to light, because the increased temperature of the bulb resulted in a dissipation of the energy (the air surrounding the bulb heated up as well).

Teachers also explored the interactions of ultraviolet (UV) light with UV-reactive beads, electricity flow with Snap Circuits, warming of model lizards (bottles filled with water) with heat lamps, and their own personal system (anything in their lives with which they were familiar, such as a coffeepot, a car, or a garden). Because the energy concept is rooted in phenomena and can be used to better understand real systems, the Energy I course focus on phenomena helped teachers view new systems from an energy perspective.

An Energy Wall to Emphasize Energy as an Analytical Tool

The physical attributes of objects and systems can be measured in different units. To measure distance, we use centimeters, meters, and kilometers when appropriate, but we also use inches, miles, fathoms, and light-years. To understand any situation where different units are used requires an ability to convert from one system of units to another. If someone is only familiar with Fahrenheit and travels to a place where the temperature in the weather report is given in Celsius, he or she may not know whether a jacket is necessary. Memorizing equations that convert one unit to another unit or using a convert app on a cell phone can help, but it is more efficient to have a sense of the meaning of someone's statement as they say it. Knowing that 0°C is equivalent to 32°F (freezing water), that 20°C is equivalent to 68°F (room temperature), and that 37°C is equivalent to 98.6°F (human body temperature) helps a person have a sense of any Celsius temperature.

While temperature is most often measured in Celsius and Fahrenheit, energy measurements are commonly given in joules (J), calories, Calories (the unit used on food labels—the capital "C" is used to mean kilocalories in this case), ergs, British thermal units (Btu), electron volts (eV), and gallons of gasoline. When we quantify temperatures, most measurements fall between –273°C (just above absolute zero) and 5000°C (the Sun's surface temperature). The numerical range of energy values that we deal with is much, much larger.

To describe a range of 1 billionth of a joule to 1 billion joules, it is useful to use exponential notation (10^{-9} J to 10^9 J). For some of the teachers in the Energy I course, exponential notation is a new and unfamiliar idea. An Energy Wall (Figure 9.1) was used throughout the course to emphasize that energy can be used as an analytical tool and to remind teachers that all energy is the same energy, but it just might be measured in different units in different situations.

Figure 9.1. Energy Wall used in the BEST Energy I course

The Energy Wall was a 15-foot-wide poster covering 18 orders of magnitude (10^{-9} to 10^9 joules) with three parallel scales (joules, gallons of gasoline, and food Calories). Teachers taking the course could place on the wall cards with the energy of the various phenomena that were discussed and make comparisons with common units of all the other phenomena that were explored. How many times would you have to lift a ball until its gravitational energy change is equal to the Calories in one cracker? How many gallons of gasoline are equivalent to the energy needs of a typical human in one year? How many butterflies might it take to power a lightbulb? The Energy Wall was a simple mechanism to present energy ideas across disciplines and contexts and to reaffirm that all energy is fundamentally the same thing.

With the renewed emphasis on engineering in the *Framework* and the *NGSS*, the Energy Wall can be a valuable means of helping teachers and students connect the gallons of gasoline that they are very familiar with and the energy calculations that use different units in classroom experiments. The importance of recognizing the magnitude of energy is a crucially important tool when discussing energy as a resource and how to best use the energy resources available to us.

Focus on Systems: An Introduction to Real-World Phenomena

Energy I was organized around a theme of energy in systems throughout the course. Systems thinking (Systems and System Models is another CCC in the *Framework* and the *NGSS*) is critical to developing a full understanding of energy transfer between objects and organisms, and across the boundaries of real-world phenomena. Systems are made of subsystems and components and can be analyzed by examining the energy transfers throughout. As a culminating project, each teacher designed one energy subsystem and then got together in groups with other teachers to link the subsystems to form a Rube Goldberg system. In a Rube Goldberg device, a simple task is completed through a roundabout path of often-unrelated tasks. For example, the task of turning the page of a newspaper can begin by putting a cup of coffee on the table, which dislodges a ball that goes down a ramp that, in turn, moves a lever that will release a switch that will turn on a light that will increase in temperature and cause a string to break that will move some pulleys that will turn the page of the newspaper.

The idea of activation energy is crucial to explaining the mechanics of the Rube Goldberg system. Activation energy is the energy input required to initiate an energy transformation or transfer process—for example, heating to initiate the release of energy through burning or pushing a book off the edge of a desk. When a Rube Goldberg system is set up, some objects are placed at heights so that they can fall and compress some springs. Both of these increase the potential energy of the system. When a switch is moved—a small activation energy—the potential energy of the system is released.

In the Energy I course, teachers typically worked in groups of four to six and were often highly engaged and enthusiastic about using their own materials and ideas to create their Rube Goldberg system. Through their spirited investigations, teachers gained experience identifying the boundaries around systems and subsystems and using the idea of activation energy to release potential energy within this complex set of interactions.

Overall, the Energy I course proved successful because of the integration of the key features of effective energy PD and multiple innovations, as well as dedication by instructors and students to the idea of using energy to cross the disciplines.

Boston Energy in Science Teaching: Energy II

Energy II, which was also team-taught by professors in different disciplines and a middle school teacher, was based on what the participants were teaching in the classroom and provided a forum for participants and instructors to provide feedback on how to use energy connections to integrate student learning both horizontally across all disciplines and vertically across all grade bands. To experience the classrooms of a great diversity of teachers in Energy II, participants videotaped their own lessons and these videotapes were used in the course. Four different Collaborative Coaching and Learning in Science (CCLS) cycles (described in more detail later) were used to provide feedback on four different lesson plans in each teacher's classroom. In each CCLS cycle, one presenter offered a 15-minute video of some of his or her classroom teaching, and three or four teachers and one professor would debrief the video using a strict protocol of clarifying questions, warm (areas of strength) and cool (areas for growth) feedback, and missed opportunities for connections. Teachers were also required to share student work from the lesson. In this way, each teacher was thinking about making energy connections in his or her authentic classroom teaching 12–16 times in the course. This practice increased the ability of the participating teachers to make connections across curriculum and grade bands and help participants make connections to how they were thinking.

In addition, Energy II used reflections and real-world problems (using energy to explain everyday situations) to enhance the CCLS learning. Energy II discussions revealed a new and novel way to teach content. In Energy I, a set curriculum was in place and the professors would present material and involve teachers in learning. In Energy II, discussions were initiated around the lessons that the teachers were implementing in their own classrooms. Small groups of teachers would have these discussions with one professor in each group. During the "missed opportunities to discuss energy in a specific lesson" discussions and the viewing of student work, questions would emerge revealing both to the teachers and the professor that there may be a need for teachers to engage in additional learning about energy. This, then, became an opportunity either to discuss that content at that time or to choose that content as the focus of a follow-up class. For example, all teachers in one section of Energy II were aware that the temperature of a room is a measure of the average kinetic energy of the molecules (primarily nitrogen and oxygen) in the air. When asked, "How fast do you think that the molecules are moving?" some of the teachers had no idea of the value and no idea of how you find out that value. This led to discussions about the gas laws and thermodynamics, including references to the Maxwell-Boltzmann distribution (which gives an expected distribution of molecular speeds based on the temperature of a substance), frequency of collisions between molecules, and time between molecular collisions. The resulting speed of approximately 1,000 miles per hour was quite surprising for some. Although the mathematics was challenging and beyond the scope of what the teachers can use in their classrooms, the discussion helped strengthen their understanding of temperature.

BEST Energy Institute

To identify how energy is taught throughout the Boston Public Schools curriculum across disciplines and grade bands, the Energy Institute brought together 12 teachers (one representing each grade) to identify the district-mandated curriculum lessons that they teach, which contain some energy learning, and to identify connections among these lessons. Over three days, 105 of the 306 grades 1–12 curriculum lessons in the Boston Public Schools curriculum were identified as relating to energy and 162 connections were described among them. For example, in fourth grade, students made cars driven by rubber bands. In sixth grade, students burned a marshmallow and a sunflower seed to release energy from food. Teachers discovered the connection between these two activities: Both involved releasing potential energy—in one system (the fourth-grade activity) from elastic energy to kinetic energy, and in the other system (the sixth-grade activity) from chemical energy to heat energy.

Using social network analysis software, a map of the most connected lessons was developed, and the eight most connected ones were identified (Table 9.1, pp. 184–185; Chen et al. 2014). Interestingly, while the main goal of the Energy Institute was to create the map, the largest impact was PD for the participating teachers. They reported learning a tremendous amount of energy content and conceptual understanding by examining their own curricula in depth and learning how connected student learning is among the horizontal and vertical spread of the curriculum. Energy proved to be a significant crosscutting vehicle for learning and connecting all science.

Table 9.1.

THE EIGHT MOST CONNECTED CURRICULUM UNITS IN THE BOSTON PUBLIC SCHOOLS CURRICULUM, WITH DESCRIPTIONS, THEMES, AND CONCEPTS, AS DETERMINED BY TEACHERS IN AUGUST 2012

Curriculum unit	Grade	Discipline	Activity description	Energy theme	Energy concept
Active Physics Home—1: Designing the Universal Dwelling	9	Physics	Students build a model home and study heat transfers.	Resources	The Sun is a source of solar energy and heat, which can be transferred by conduction, convection, and radiation.
FOSS Populations and Ecosystems—5: Finding the Energy	8	Biology	Students burn a cheese ball and record the temperature change of a cup of water above the flame.	Conservation	Energy stored in food can be measured. Energy is conserved as it is transformed from one form to another.
FOSS Diversity of Life—5: Seeds of Life	7	Biology	Students germinate seeds in light and dark.	Forms and Transformations	Seeds store energy that can be used to germinate.
Living by Chemistry Fire—1: Evidence of Change	10	Chemistry	Students mix materials of different temperatures.	Forms and transformations	Energy is conserved and tends to disperse.
Living by Chemistry Water—3: Water Vapor	3	Chemistry	Students observe two cups with melting ice or soaked paper towels and record evidence of phase changes.	Forms and transformations	Energy is required for phase changes.

Table 9.1 (*continued*)

Curriculum unit	Grade	Discipline	Activity description	Energy theme	Energy concept
Biology—A Human Approach: Matter, Energy, and Organization—8: The Cellular Basis of Activity	11	Biology	Students examine photosynthesis through an experiment and cellular respiration by designing a marathoner's snack.	Forms and transformations	Energy is stored in matter.
FOSS Weather and Water—5: Convection	6	Earth science	Students layer different temperature waters in a vial and create convection by heating.	Systems	Temperature affects the density of air and water. Hot air rises. Temperature differences can transfer energy through convection.
FOSS Chemical Interactions—4: Kinetic Energy	8	Chemistry	Students make a simple thermometer with water and observe expansion with heating.	Systems	Matter expands when kinetic energy increases.

Source: Adapted from Chen, R. F., A. Scheff, E. Fields, P. Pelletier, and R. Faux. 2014. Mapping energy in the Boston Public Schools curriculum. In *Teaching and Learning of Energy in K–12 Education,* ed. R.F. Chen, A. Eisenkraft, D. Fortus, J. Krajcik, K. Neumann, J. Nordine, and A. Scheff, pp. 135–152. New York: Springer International Publishing. Used with permission.

Note: FOSS = Full Option Science System. FOSS Third Edition: Energy and Electromagnetism module, 2012. Developed at the Lawrence Hall of Science and published and distributed by School Specialty Science/Delta Education. Copyright © The Regents of the University of California.

The Energy Project at Seattle Pacific University

The Energy Project is a National Science Foundation–funded project at Seattle Pacific University (*www.energyprojectresources.org*). The goal of the project is to learn how to best teach energy in K–12 classrooms. Through a variety of PD opportunities for grades 4–5 teachers (Understanding Energy 1, Understanding Energy 2), secondary teachers (Energy 1, Energy 2), and inservice K–12 teachers (Teaching Seminar), the project has discovered several successful PD strategies.

One such strategy is the Energy Theater, in which teachers act as units of energy and focus on how energy is conserved as it is transferred from system to system. Energy cubes, which are six-sided dice with a different energy form (e.g., thermal energy, kinetic energy) on each side, are used to represent energy transformations and transfers in phenomena. In addition to providing a way to easily represent energy transformation and transfers, the cubes help teachers develop an intuition for the conservation of energy because they must consider what will become of each energy cube that they use to represent a phenomenon or process. Teachers also develop "energy tracking diagrams," which are teacher-invented representations of energy transformation or transfer processes that were found to be effective analytical tools for examining real-world systems.

By employing a variety of strategies throughout all the PD opportunities, teachers learn to use multiple models to represent energy flows through different systems. In addition, the Energy Project focuses on highlighting the difference between "people talk," the language people use in everyday situations, and "feature talk," the language used for describing specific scientific phenomena. By delineating these two types of talk, the PD helps teachers to draw a conceptual boundary between energy-related words when they are used in everyday versus scientific context.

A critical feature of the Energy Project is that all of the strategies the teachers use in their PD are easily transferable to the classroom. Thus, teachers must only adapt, rather than invent, representations to help their own students explore the role of energy in phenomena. The Energy Project helps teachers create a bridge between their own content learning and their ability to support students in their own classrooms.

"Connected Science: Powering Our Lives" Course at Trinity University

At Trinity University in San Antonio, Texas, an interdisciplinary group of university scientists, a university-based science educator, and several local teachers collaborated to develop a new science content course for future K–8 teachers called Connected Science: Powering Our Lives. Content investigations in this course were organized around two project-based science modules that used the driving questions "How do we power our cars?" and "How do we power our bodies?" By exploring uses of fuel and food, future teachers learned to use the same representations of energy across multiple systems, and they learned core scientific ideas while engaging in scientific practices (such as asking questions and constructing scientific explanations) as they designed and conducted investigations. In this way, future teachers engaged in the type of three-dimensional learning (NRC 2014) that the *Framework* and the *NGSS* emphasize.

The Connected Science course is taught by the university-based science educator, and laboratory investigations were designed to use the types of materials that teachers were likely to find in K–8 classrooms. By using such materials and emphasizing energy as a CCC in concert with investigations aligned to a range of DCIs, the Connected Science course models concrete strategies and conceptual tools for engaging young learners in three-dimensional learning.

Future K–8 teachers who have participated in the course have commented that it helped them learn more about core scientific ideas, the scientific process, and why energy is a powerful scientific concept. They also reported feeling more confident in their ability to guide young students in learning about science.

District-Level Professional Development

In addition to formal courses at the inservice and preservice levels, PD for energy learning and implementation can be conducted during afternoon, full-day, or longer workshops. In each case, it is important to consider the imperatives, constraints, and opportunities of and for focusing on energy as a DCI, CCC, and scientific practice. Naturally, vertical integration and interdisciplinary collaboration will enhance any PD workshop. Further, basing the learning on the teachers' experiences, classroom activities, and appropriate *NGSS* standards will help teachers see the connections and appreciate the strength of the energy concept and the Five Big Ideas about energy.

The strategies identified in the following sections are examples of structures for supporting professional collaboration around the CCC of energy. Like any high-quality PD initiative, these strategies are most effective when they are included as a part of a sustained professional support program for teachers (Loucks-Horsley et al. 2010).

Vertical Teaming

In a half-day or full-day workshop format, teacher-participants from each grade band (elementary, middle, high school) examine a high-level problem related to energy (an AP exam question from biology, chemistry, physics, or environmental science works well—particularly one that includes a laboratory focus). They first go over the solution to the problem, and, in the process, teachers focus on scientific ideas in the problem and clarify their understanding of these ideas. They then discuss what students must know and be able to do to solve the problem. This "unpacking" of the knowledge required to solve the problem reveals that some of the knowledge and skills are built in elementary classrooms whereas other content and practices are introduced in middle school or later. This process can lead to concrete discussions about how each grade band contributes key experiences for students as they build increasingly sophisticated understandings of energy over time. By referring to the *NGSS*, teachers can explore how these standards help students develop ideas about energy forms, transformations, transfers, conservation, and dissipation over the course of years.

This process works best when teachers discuss not just content-focused ideas but also how teachers in each grade contribute to an increasingly robust evidence base for introducing core ideas. That is, the PD facilitator should prompt teachers to consider not only when ideas are introduced, but how. The more instruction incorporates practices, DCIs, and CCCs, the better positioned students are to develop a robust understanding that promotes deeper learning over time.

Collaborative Coaching and Learning in Science

In the CCLS model, teams of four to eight teachers (from either the same grade band or across grade bands) come together to form a professional learning community (PLC). In this type of PLC, teachers from any grade observe the others teach (through in-person classroom observations or sharing a video of the participant teaching). By following a protocol (McDonald et al. 2003), debriefing discussions can be focused tightly on a set of targeted prompts, which typically increases the impact of the conversations and enhances the efficiency of meeting time.

The process begins by participants viewing (at the beginning of the meeting or at home prior to the meeting) a video of one of the participants teaching. Then, participants ask clarifying questions for about 2 minutes and provide warm and cool feedback about what they saw for about 10 minutes. In this process, the teacher who is being observed is silent and listening. After this general discussion, all participants are invited to a roughly 8-minute discussion about the energy connections within the lesson. These connections may have been observed in the lesson, or they may be missed opportunities that an observing teacher recognized while watching the lesson. After this discussion, all participants explore samples of student work for about 5 minutes to consider evidence of student learning about energy. In the final 5 minutes, all teachers are invited to reflect on their learning during their discussion, with the observed teacher speaking last. It is helpful in this process for one teacher to act as facilitator to ensure that the group sticks to the time limits for each phase of the protocol-driven discussion.

After the roughly 30-minute debriefing discussion, teachers will have considered key science content ideas about energy, opportunities to make new connections to the energy concept, and evidence of student learning. Over time, this process can help a group of collaborating teachers come to a shared understanding of how to focus energy instruction on the most important ideas (i.e., the Five Big Ideas about energy) and to support student learning over time.

School-University Partnerships

School districts can benefit from partnerships with university faculty and vice versa. By involving content and concept experts from higher education, the richness of the discussions and the depth of the content presented in district-based PD are often extended. Further, such partnerships provide university-based science faculty with opportunities to engage with local science teachers and students. PD facilitators may ask the invited university faculty to share their perspective on energy and how they use it within their discipline(s). By involving university faculty with K–12 faculty in side-by-side discussions about how energy is taught in K–12 schools, both groups can develop a clearer picture of how the Five Big Ideas about energy are used within each discipline and how students can be supported in developing sophisticated understandings that serve them well both in K–12 education and beyond. Further, involving content experts from the university typically helps illuminate the tentative nature of scientific knowledge while reaffirming that some scientific ideas (such as energy conservation) are so extensively tested that they are no longer the subject of significant investigation.

Involving science experts can be tricky because many university science faculty were never taught explicitly how to teach. Thus, the PD facilitator must take special care to structure discussions and activities such that they present teachers with insights that are useful to their classroom teaching. One way to structure such discussions is to ask an interdisciplinary group of scientists to serve on a panel to discuss how they use energy in their work (virtually every scientist does). After seeding the discussion with introductory questions (e.g., "How do you use energy in your work?" "What should K–12 students learn about energy?" "Why is energy so important in our world?"), the facilitator can encourage teachers to ask their own questions. Further, by providing scientists with information about the *Framework*, the *NGSS*, or the importance of the Five Big Ideas about energy in K–12 education ahead of the discussion, the facilitator can help ensure that teachers will find the experience to be both clarifying and reaffirming.

Summary

Energy has always been an important area for science teacher PD because the concept is so central to science. Yet, PD experiences for energy are often discipline based or otherwise fail to emphasize the crosscutting nature of the energy concept. Because of its position in the *NGSS* as both a DCI and a CCC, energy can be used in PD for discussing broader issues related to teaching science according to the three-dimensional learning that the *NGSS* emphasize.

Energy PD can promote discussions of the role of language in science, strategies for connecting in-school and out-of-school learning, and the use of models in representing scientific phenomena, to name a few. Further, because energy is so central to each science discipline and a concept that is built during the years of K–12 science instruction, energy PD can be structured to bring teachers together around a common instructional topic to do both horizontal and vertical collaboration. The Five Big Ideas about energy can help frame energy PD and provide a common language to help teachers understand how each idea is connected and how students can be supported in developing more sophisticated understandings over time.

Effective energy PD is based on real-world phenomena, connects grade bands and disciplines, emphasizes the use of energy as an analytical tool, and integrates three-dimensional learning. Although other features can make PD effective as well, these four key features are particularly effective when teaching teachers how to connect curricula through energy.

Universities, school districts, and PD providers all have a role to play in supporting teachers' understanding of the *NGSS* and how to implement them in real classrooms. Yet, the *NGSS* contain a lot of ideas that can be overwhelming to consider all at once. Energy PD may be useful for providing a smaller chunk of content exploration that will lead into a deeper understanding of how a DCI and a CCC are different but complementary and how scientific practices help learners explore topics that are relevant in school and their everyday lives.

References

Chabalengula, V. M., M. Sanders, and F. Mumba. 2012. Diagnosing students' understanding of energy and its related concepts in biological context. *International Journal of Science and Mathematics Education* 10 (2): 241–266. Also available online at *http://doi.org/10.1007/s10763-011-9291-2*.

Chen, R. F., A. Eisenkraft, D. Fortus, J. S. Krajcik, K. Neumann, J. C. Nordine, and A. Scheff, eds. 2014. *Teaching and Learning of Energy in K–12 Education*. New York: Springer.

Intergovernmental Panel on Climate Change. 2000. *Special report on emissions scenarios: A special report of Working Group III of the Intergovernmental Panel on Climate Change*, eds. N. Nakićenović and R. Swart. Cambridge, UK: Cambridge University Press.

Loucks-Horsley, S., K. E. Stiles, S. E. Mundry, N. B. Love, and P. W. Hewson. 2010. *Designing professional development for teachers of science and mathematics*. 3rd ed., expanded ed. Thousand Oaks, CA: Corwin Press.

McDonald, J. P., N. Mohr, A. Dichter, and E. McDonald. 2003. *The power of protocols: An educator's guide to better practice*. New York: Teachers College Press.

National Research Council (NRC). 2012. *A framework for K–12 science education: Practices, crosscutting concepts, and core ideas*. Washington, DC: National Academies Press.

National Research Council (NRC). 2014. *Developing assessments for the Next Generation Science Standards*. Washington, DC: National Academies Press.

NGSS Lead States. 2013. *Next Generation Science Standards: For states, by states*. Washington, DC: National Academies Press. *www.nextgenscience.org/next-generation-science-standards*.

CHAPTER 10

SYSTEMIC SUPPORTS FOR TEACHING ENERGY AS A CROSSCUTTING CONCEPT

PAMELA PELLETIER AND ALLISON SCHEFF

The vision described in *A Framework for K–12 Science Education* (NRC 2012) is bold. Though built on decades of research in learning and science education, there are many recommendations in the *Framework* that are unfamiliar to teachers and suggest a substantial change in how we conceptualize effective science instruction. Perhaps the most important aspect of the vision described in the framework is its emphasis on *three-dimensional learning*, meaning that science instruction should consistently engage learners in science and engineering practices, focus on a small set of disciplinary core ideas (DCIs), and emphasize the importance of crosscutting concepts (CCCs) that all scientists use in their work. Although the idea of a CCC is not new (see American Association for the Advancement of Science 1990, 1993), the *Framework* includes a new emphasis on its importance in science instruction; also, the *Next Generation Science Standards* (*NGSS;* NGSS Lead States 2013) emphasizes at least one CCC (along with a DCI and a practice) in every performance expectation. Both the *Framework* and the *NGSS* include a much more explicit focus on CCCs in science instruction and assessment. Supporting teachers in realizing this vision requires a systemic focus on the teaching and learning process.

This book focuses on one of the most important ideas within the *Framework* and the *NGSS:* energy. As both a DCI and a CCC, energy plays a central role in the *NGSS*, and it is this special role that both demands special attention and serves as an opportunity to explore the difference between DCIs and CCCs in science and ways to teach them. The first nine chapters of this book have explored the benefits and challenges associated with presenting energy as a CCC and have focused on using the Five Big Ideas about energy (see Chapters 1 and 2) to guide energy instruction, assessment, and professional development (PD).

In this chapter, we zoom out from a classroom-centric view to consider a systemic perspective on supporting teachers in the teaching of CCCs in general and the energy concept in particular.

Although the discussion in this chapter focuses on supporting energy as a CCC, you will notice that very few of the issues discussed below are specific to energy. That is, the practice of unpacking energy as a CCC and considering how to systematically support teachers in emphasizing this CCC in their instruction serves as a model for thinking about other CCCs as well.

We hope this chapter will be informative for teachers, but this chapter is written with a broader set of stakeholders in mind. School and district leaders, parents, business and industry, and higher education faculty all have a role to play in supporting teachers to reshape their instruction to emphasize the new vision in the *Framework* and the *NGSS*. By focusing on energy as a CCC, we hope that this chapter serves as a case study for providing a broader set of supports for teachers.

In this chapter, we address four major questions:

1. What is gained by emphasizing energy as a crosscutting concept?

2. What are some challenges in emphasizing energy as a crosscutting concept?

3. How do we address the challenges in emphasizing energy as a crosscutting concept?

4. Who has a role to play in supporting teaching that emphasizes energy as a crosscutting concept?

What Is Gained by Emphasizing Energy as a Crosscutting Concept?

The *NGSS* hold the promise of promoting greater coherence in science instruction within disciplines, across disciplines, and over time. CCCs are a critical piece of this coherence because they emphasize the development of a consistent perspective and analytical tools as students build competencies in science. Yet, many teachers already struggle with teaching in ways that simultaneously develop both science content (DCIs) and practices; the addition of CCCs into the mix may feel simply overwhelming to many teachers.

If done well, an emphasis on CCCs should clarify rather than complicate science instruction. CCCs can help students better understand key content, analytical tools, and connections across the disciplines, providing additional access points or touchstones to meaningful learning. CCCs provide an organizational framework to connect important, enduring, and unifying scientific ideas across individual disciplines like biology and physics (NRC 2012).

Energy is an idea that is used by every science discipline, but the tools for using energy can look very different in each. While an expert may easily see commonalities among the variety of disciplinary energy tools, learners often struggle with seeing connections or may even think that the energy they learn about in biology is somehow fundamentally different than the energy they learn about in physics class. Learners need explicit support in making connections across contexts (Commission on Behavioral and Social Sciences and Education [CBASSE] 2000). The emphasis on the aspects of energy that make it crosscutting is important for helping learners develop a set of consistent and useful ideas about energy.

Of course, energy is not just important across the scientific disciplines—it plays a central role in our everyday lives as well. Outside of science class, we regularly encounter energy in the context of nutrition, electricity use, efficient automobiles, global warming, and even our personal sense of vigor and vitality. Energy makes connections not only across the sciences but also to other academic disciplines and to our everyday lives (see Figure 10.1).

Figure 10.1. Energy is important in a variety of areas because the disciplinary core ideas about energy are also crosscutting.

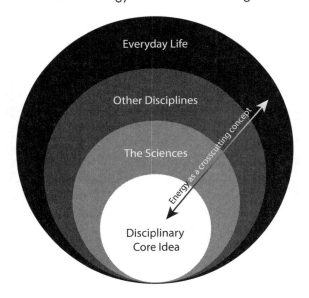

Students often learn about energy in disconnected ways, even within one science discipline. For example, biology students may study photosynthesis in one unit of study and then begin a unit on ecosystems. Of course, energy is a critical feature of both units, but the way in which energy ideas are presented is often different. Without making the connections across units explicit for students, they are unlikely to see these connections on their own. Although it may be hard to believe that students could learn about energy in the photosynthesis reaction and not see the connection between this energy and that which limits populations in ecosystems through food scarcity, study after study has confirmed that students have trouble making such conceptual links across contexts, even when they are from the same discipline (Lave and Wenger 1992; Mestre 2005). The unintentional separations of energy ideas across instructional units make it harder for students to develop an integrated understanding of energy within the context of a discipline, let alone across disciplinary boundaries.

One or more of the Five Big Ideas about energy (i.e., forms, transformation, transfer, conservation, and dissipation) are applicable in every situation in which energy is used to make sense of natural and designed systems. Thus, framing energy phenomena in terms of the Five Big Ideas can help students form conceptual bridges across contexts. These Big Ideas also give students a common language to consistently discuss energy across disciplinary boundaries (see Chapter 4). As students learn to connect a larger body of knowledge around a small set of Big Ideas, their understanding becomes more integrated (Linn et al. 2004), and the better-integrated students' ideas become, the more likely they are to be able to make sense of new situations and to use their understanding to guide future learning (diSessa and Wagner 2005; Linn and Eylon 2000; Nordine, Krajcik, and Fortus 2011). Instruction that emphasizes the crosscutting nature of energy is key to helping students develop a deep understanding of energy throughout K–12 education and apply this understanding to make sense of energy-related situations that they will encounter throughout their lives.

What Are Some Challenges in Emphasizing Energy as a Crosscutting Concept?

Most of us were not taught explicitly about energy as a CCC during our formal education. Likewise, teacher training typically has not included an explicit focus on teaching with a set of specific CCCs in mind. A primary challenge associated with emphasizing energy as a CCC is that the idea is new to many in today's teacher workforce. This novelty manifests in challenges associated with teachers' day-to-day routines, the nature of assessments, and available curriculum supports.

Taking on Something New

Reshaping instruction is not easy. The Concerns-Based Adoption Model can be a useful lens through which to view the process by which new innovations in teaching take hold (Loucks-Horsley 1996). This model stresses that teachers who are faced with a new recommendation for teaching will typically be concerned with questions such as "How does it affect me?" It is often not clear to teachers how a set of new recommendations will affect their planning, teaching, and assessment routines. Yet, we often push for teachers to implement new ideas without adequate attention to the personal and professional challenges associated with making such changes in their instruction.

Any innovation in science teaching and learning will fail if teachers do not see the value in implementing it. Thus, inservice PD efforts need to account for the difficulties that are inherent in adopting new innovations, particularly in settings where teachers are constantly being asked to implement the latest thing to maximize student performance. To truly have an impact on teacher practice and student learning, teacher leaders must be realistic about the challenges associated with reshaping one's instructional practice.

Assessment

Like it or not, high-stakes assessments drive many of the decisions that are being made in schools and districts today, and teachers are critically concerned about what it may mean for their students if they do not perform well on these standardized tests. This presents teachers with yet another dilemma that requires finding the balance between what is important to learn and what may be on the test. For students and teachers to see the importance of energy as a CCC, assessments must emphasize the crosscutting nature of energy (see Chapter 8).

The NRC Committee on Developing Assessments of Science Proficiency in K–12 (NRC 2014) argues that states' approaches to science assessment does not align with the goals of the *NGSS* because they fail to track students' gradual progress toward learning goals over time. Furthermore, the NRC committee makes the case that current large-scale assessments do not adequately require students to demonstrate knowledge of how scientific ideas and practices are integrated. Without an assessment system that is more closely aligned with teaching with CCCs in mind, lasting instructional changes are unlikely to take hold.

Curriculum

There are few examples of curriculum materials that emphasize energy as a CCC, or CCCs more generally. Without access to such materials, teachers are in a difficult situation in which they are asked to reshape their own instruction without sufficient resources and support. Even if teachers accept that teaching energy as a CCC is a valuable instructional objective, they face the additional burden of needing to seek out the materials to support this type of instruction or to work on their own to develop them.

Traditional textbooks have long been criticized for a lack of instructional coherence and inadequate focus on the big ideas of science (Fortus and Krajcik 2012; Kali, Linn, and Roseman 2008; Kesidou and Roseman 2002). Without more powerful resources for teaching with three-dimensional learning in mind, teachers will struggle with reshaping their instruction.

How Do We Address the Challenges in Emphasizing Energy as a Crosscutting Concept?

All science teachers are embedded within a broader educational system that extends well beyond the walls of their classrooms. Thus, addressing the challenges associated with reshaping science instruction will involve efforts from a wide range of influences. In this section, we discuss the role of inservice and preservice teacher training, assessments, and curriculum materials in supporting science teachers to teach with CCCs such as energy in mind.

Inservice Professional Development

The transition from one set of standards to another is not easy, even if the standards contain powerful new ideas that teachers recognize as important. In addition to the typical challenges that come with adapting to a new set of standards, teaching with CCCs in mind requires teachers to become more connected to content outside of their home disciplines. Teachers need significant, ongoing PD with peers and experts to effect such instructional changes.

Too often, PD in science consists of one-shot workshops that are not closely aligned with intended content learning goals, are not sustained over time, and fail to recognize and account for key problems of practice that teachers encounter in their work (Loucks-Horsley et al. 2010). Without such features, PD initiatives are unlikely to be successful (Guskey and Yoon 2009; Ingvarson, Meiers, and Beavis 2005).

To be effective, teachers will need sustained opportunities to bolster their content knowledge outside their own discipline and training in pedagogical content supports that stress how to connect to the Five Big Ideas about energy in their own classroom. It is not enough for PD to identify opportunities for making energy connections across disciplines; professional developers must carefully and deliberately support teachers in both identifying substantive connections and developing a set of strategies for using energy consistently across contexts and disciplines. It is the quality of the connections, not the quantity of them. Educators need guidance and practice to learn where these connections are best made and how to incorporate these connections into their lessons. The Five Big Ideas about energy provide a consistent perspective from which to evaluate cross-disciplinary connections and develop pedagogical supports for student learning.

Chapter 9 presents the example of the Boston Energy in Science Teaching (BEST) project, a PD effort aimed at increasing teachers' ability and comfort in using energy as a CCC. This project serves as a model for involving teachers in reconceptualizing their instruction to make the crosscutting nature of energy more explicit by identifying and evaluating energy connections within their existing curricula. Chapter 9 also discusses the Energy Project at Seattle Pacific University, which focuses on providing teachers with a set of conceptual tools and representations for using energy as a CCC with their students. Findings and tools from both of these projects can inform local workshops by providing a starting place for professional developers.

Preservice Teacher Education

University-based schools of education have a tremendous opportunity to reshape the teaching landscape. The teacher educators in these institutions have a unique opportunity to introduce preservice teachers to a set of pedagogical tools and strategies for teaching with CCCs in mind. By placing a strong emphasis on planning for three-dimensional learning, education faculty can help preservice teachers emphasize both scientific practices and CCCs as they develop their competency in lesson planning.

In addition to supporting their three-dimensional lesson planning, collaboration between education faculty and science faculty may strengthen preservice teachers' understanding of the DCIs in the *NGSS* and their ability to connect seemingly disparate ideas through energy. Science content courses, particularly those aimed at future K–8 teachers, should exemplify three-dimensional learning and model the kind of pedagogical supports needed for teaching with CCCs in mind. For future teachers to effectively make connections across disciplines, they need to develop a rich understanding of how scientific ways of thinking are consistently applied across a wide variety of natural phenomena and designed devices. The *Framework* summarizes why CCCs are essential to one's ability to understand science and represent it to learners:

> [Crosscutting] concepts help provide students with an organizational framework for connecting knowledge from the various disciplines into a coherent and scientifically based view of the world. … These concepts should become common and familiar touchstones across the disciplines and grade levels. Explicit reference to the concepts, as well as their emergence in multiple disciplinary contexts, can help students develop a cumulative, coherent, and usable understanding of science and engineering. (NRC 2012, p. 83)

In conjunction with emphasizing CCCs in science and education courses for preservice teachers, university-based faculty must provide future teachers with opportunities to develop their pedagogical content knowledge through structured interactions with real K–12 students in real classrooms.

Universities should give serious consideration to offering science content courses specifically for preservice elementary teachers that are organized around CCCs such as energy and emphasize three-dimensional learning as described in the *Framework* (see Chapter 9). By providing elementary teachers with the opportunity to fulfill their science requirements with such a course instead of typical courses for nonmajors (which are, too often, uninspiring for students and faculty alike), science and education faculty can provide a stronger base in the foundational scientific ideas for teachers who will be responsible for a child's first eight or nine years of science learning.

Assessment

The NRC report *Developing Assessments for the* Next Generation Science Standards (2014), stresses that both formative and summative assessments should be designed to reflect the importance of three-dimensional learning by simultaneously emphasizing practices, DCIs, and CCCs. To more fully support instruction that emphasizes the crosscutting nature of energy, assessments must demonstrate that such an understanding is valuable.

Too many assessments today (both large scale and classroom based) are focused on measuring isolated facts and skills without sufficient attention to whether and how learners can connect ideas to one another to make sense of meaningful situations. As discussed in Chapter 8, high-quality assessments that emphasize the crosscutting nature of energy begin in the classroom, because it is these assessments that most strongly signal to learners what knowledge and abilities are most important. Teachers have a critical role to play in ensuring that their own formative and summative assessments exhibit the principles of three-dimensional learning by including a consistent emphasis on practices, DCIs, and CCCs.

In their 2014 report, the NRC Committee on Developing Assessments of Science Proficiency in K–12 called for a new generation of large-scale assessments that are gradually introduced, are fair to diverse learners, and leverage new technologies for assessing a large number of students in meaningful ways (NRC 2014). Over time, such a system would both give a more accurate picture of student understanding and drive instructional practices that emphasize three-dimensional learning.

Curriculum

Lessons enacted in science classrooms must make much more explicit connections to the crosscutting nature of energy. Yet, teachers often work from materials that were written before the *Framework* or that simply do not include an explicit focus on CCCs such as energy. Teachers in this situation must thoughtfully adapt the materials they are provided to emphasize energy in their instruction.

When looking for places to emphasize energy in existing curriculum materials, the rule is quality over quantity. Energy is such a far-reaching scientific idea that it can be connected to just about any scientific phenomenon, but that does not mean that making every connection is beneficial to learners. In fact, if there are too many connections to energy that occur too frequently, students may begin to perceive this as just another layer of "noise" within science lessons (Levy et al. 2013). Some topics are powerful opportunities to build students' crosscutting understanding of energy, while others are less so. This book has focused on three DCIs from life, physical, and Earth science, as follows:

1. Many organisms use the energy from light to make sugars (food) from carbon dioxide from the atmosphere and water through the process of photosynthesis,

which also releases oxygen. In most animals and plants, oxygen reacts with carbon-containing molecules (sugars) to provide energy and produce waste carbon dioxide. (LS1.C)

2. That there is a single quantity called energy is due to the remarkable fact that a system's total energy is conserved as smaller quantities of energy are transferred between subsystems—or into and out of the system through diverse mechanisms and stored in various ways. (PS3.A)

3. All materials, energy, and fuels that humans use are derived from natural sources, and their use affects the environment in multiple ways. Some resources are renewable over time, and others are not. (ESS3.A)

We selected these DCIs because they represent high-leverage opportunities to empha-size the crosscutting nature of energy. Although we can certainly layer an energy focus onto instruction related to biodiversity, wave refraction, or volcanism, the learning goals listed above are opportunities to help students understand how the energy in biology, the energy in physics, and the energy in Earth science are all fundamentally the same and adhere to the same five rules (the Big Ideas).

Being thoughtful about where to emphasize energy connections is similar to supporting students' science and engineering practices. In teaching science through inquiry, we do not ask students to design, conduct, carry out, and present investigations every day in class. Doing so would be frustrating for students and detract from the main idea of the lesson. Likewise, emphasizing energy too frequently within situations that are not high-leverage will only serve to confuse and complicate the main ideas of science lessons and likely frus-trate students. Teaching energy as a CCC does not mean that every place that could have an energy connection should have an energy connection.

Teachers should keep a watchful eye out for opportunities within their curriculum to support students in understanding the crosscutting nature of energy by connecting to the Five Big Ideas about energy. If using some or all of the Five Big Ideas helps students make sense of the content of the lesson in a way that deepens understanding or develops con-nections across disciplines, then do it. If, however, making an energy connection means adding another layer of complexity to a lesson, then it is probably not a high-leverage opportunity for learning about the crosscutting nature of energy.

Teacher curriculum adaptations are a critical part of effective science instruction, but teachers can only adapt so much before the process becomes overwhelming and impracti-cal. Teachers need access to materials that are designed to emphasize CCCs such as energy. With the *Framework* and the *NGSS*, curriculum developers and textbook publishers have an opportunity to produce materials that emphasize three-dimensional learning and, as a result, provide teachers with resources for teaching energy as both a DCI and CCC.

Curriculum materials that emphasize CCCs must recognize high-leverage opportunities for teaching about energy, make connections to energy as a CCC explicit, and build students' competency over time. There is evidence that materials that are designed in this way can help students develop an understanding of energy as a CCC (Fortus et al. 2015).

Regardless of the quality of instructional materials, teachers will—and should—make adaptations to fit their students and school setting. When teachers have access to curriculum materials that exemplify three-dimensional learning, they are in a much stronger position to adapt activities for use within their local instructional context.

Who Has a Role to Play in Supporting Teaching That Emphasizes Energy as a Crosscutting Concept?

Teachers

Teachers are the single most powerful lever for instructional change in their school and the broader educational system. We hope that this book provides valuable insight for teachers into teaching energy as a CCC and, more broadly, serves as a model for thinking about how to address other CCCs as well.

Building students' understanding of energy takes years and requires connections across disciplines. No one teacher can meet this challenge alone. Even informal conversations with colleagues around the Five Big Ideas about energy can help identify opportunities for connecting energy lessons between grade levels and across disciplines, and these connections can lead to broader discussions around other CCCs and provide a vehicle for enhancing instructional coherence within a school.

Although teachers are a critical catalyst for instructional change, they are simply too busy with the day-to-day work of the classroom to do it all on their own. To promote substantive and sustainable changes in instruction that promote CCCs like energy, teachers need support from other key stakeholders.

School and District Leaders

School and district-level leaders play a critical role in supporting teachers as they emphasize CCCs such as energy in their teaching. Even though school-level administrators must dedicate much of their time to noninstructional duties, principals are the instructional leaders of their school. They set the agenda for instructional improvement and work to ensure that teachers have access to the resources they need to teach well. Effective school administrators also create a culture that enables teachers to spend their time focused on the instructional issues that matter most.

Administrators often feel tremendous pressure to ensure that their district performs well on statewide high-stakes assessments, and if the state assessment system does not align well with the recommendations of documents such as the *NGSS*—as it typically does not (NRC 2014)—it can seem unfruitful for administrators to advocate for such instructional change in their schools. However, it is critical for administrators to draw a conceptual distinction between learning and assessment practices, and the research is clear—students learn best when they work on meaningful problems that help them use a consistent set of big ideas within a variety of contexts (CBASSE 2000; Duschl, Schweingruber, and Shouse 2007). Although it may take the assessment system time to catch up, administrators play a critical role in assuring teachers that high-quality instructional practices will lead to better outcomes and that they will be supported by their school and district administrators as long as they are making sound instructional choices.

School and district administrators set the stage for teachers to collaborate with one another on meaningful problems of practice and ensure that systems are in place to allow teachers to focus intensively on high-leverage instructional issues (Breidenstein et al. 2012). A deep understanding of CCCs is not built in a single unit or even in a single year; for teachers to be able to emphasize CCCs such as energy in their instruction, they must have time to collaborate together within and between grade levels. School and district administrators are critical to ensuring that such collaborations take place.

Parents

Parents often do not realize their power to have an impact on the school or the district, and teachers may find this stakeholder group to be a great ally in advocating for the use of energy and other CCCs. There are state mandates for almost all districts in the United States that limit school district autonomy to some extent (e.g., content standards, fiscal policies), but parents can be the greatest champions for creating broad-based curricular and extracurricular initiatives to be implemented in their own school district. Although the selection of educational materials, instructional practices, and curriculum development should be left in the hands of qualified educators, parents have a strong voice in advocating for meaningful instructional change. To be strong advocates, it is important for parents to know what is being recommended at the national level by scientists, industry leaders, and educators. The *Framework* is written for a general audience and is freely available online (*www.nap.edu/ catalog/13165/a-framework-for-k-12-science-education-practices-crosscutting-concepts*), and this document would be a valuable resource for parents to strengthen their voice as they advocate for science instruction that is connected and meaningful to their children's lives both inside and outside of school. Asking how schools support students in learning about key social issues such as energy may be a good place to start.

Higher Education

Teachers tend to teach how they were taught, and, unfortunately, science instruction in higher education often bears little resemblance to the vision in the *Framework* (Savkar and Lokere 2010). Major reasons for this mismatch involve issues of culture and rewards within universities (Anderson et al. 2011) that are beyond the scope of this book, but university professors—especially those with tenure—bear ultimate responsibility for the quality and character of teaching at their institution and have almost total autonomy to implement instructional changes.

Bruce Alberts, former president of the National Academy of Sciences and editor-in-chief of *Science*, warned university science faculty that "If we continue to emphasize all of the facts of science in our college introductory courses, parents and politicians will continue to expect high-school science courses based on textbooks that convey a string of scientific facts and nothing more" (Alberts 2005, p. 740). Writing in particular about the treatment of energy in many university-based science courses, Melanie Cooper, a chemist at Michigan State University, and Michael Klymkowsky, a biologist at the University of Colorado, declare that, "Simply put, we (biologists, physicists, and chemists) are not providing a coherent pathway for most students to develop a usable understanding of energy, particularly at the atomic–molecular level. We are failing our students by not making explicit connections among the way energy is treated in physics, chemistry, and biology. We cannot hope to make energy a crosscutting idea or a unifying theme until substantive changes are made to all our curricula" (Cooper and Klymkowsky 2013, p. 309).

University faculty largely set the criteria for what counts as science and what does not. Unless university science courses—particularly those taken by future teachers—resemble the vision of science that we hope K–12 educators will convey to their own students, it will be even more difficult for university graduates to align their teaching with the vision of the *Framework*.

Aside from rethinking university-level science pedagogy, higher education faculty have a role to play in working directly with practicing teachers. Science and education faculty can collaborate to offer outreach programs that provide K–12 teachers with experiences in learning and doing science that are aligned with the *Framework*. By linking such learning experiences with explicit discussions of pedagogical supports, university faculty can support teachers in connecting their own science learning to changes in instructional practice. Offering such a PD experience focused on the concept of energy and its power as a CCC may be an excellent opportunity to engage with teachers from across a range of grade level and disciplines.

Between using university-based courses to model three-dimensional science instruction and collaborating to provide effective PD outreach experiences for teachers, higher education faculty—particularly those in the sciences—can be a powerful voice for change in K–12 science teaching.

Business and Industry

The reports *Rising Above the Gathering Storm* (Institute of Medicine, National Academy of Sciences, and National Academy of Engineering 2007) and *Rising Above the Gathering Storm, Revisited* (NRC 2010) cite economic research that attributes well over half of the growth in the United States economy during the 20th century to advancements in knowledge, particularly technology. The development of tomorrow's workforce and the maintenance of the competitive advantage that America has enjoyed throughout much of the 20th century rely on how well the K–12 education system can prepare students to function in a world increasingly influenced by scientific and technological advances.

Workforce preparation does not begin when a student receives an associate's or bachelor's degree; it begins in the K–12 system. As the world's grand challenges—many of them related to the need to harness more energy resources—become more complex, the solutions to these challenges may very well lie in thinking differently and across traditional disciplinary boundaries. The teaching and learning of energy as a CCC is a promising strategy for preparing people to better understand science content, make connections between disciplines, work together to find solutions, and know how to approach unfamiliar systems. Students who can do these things will be well positioned to contribute to continued economic growth and development. Furthermore, it is the strength of this three-dimensional approach to learning that directly supports the needs of industry. Local and national employers can be valuable partners to educators in helping to communicate how science and engineering practices, DCIs, and CCCs directly relate to high-demand workforce sectors.

Through targeted support (e.g., sponsorship, advocacy, externships) of projects such as collaborative curriculum development and PD initiatives that align with the vision in the *Framework*, business and industry can play a powerful role in supporting meaningful three-dimensional science instruction. Because of its importance as both a DCI and a CCC, sponsorship of initiatives that support energy teaching and learning may be a good place to start.

Summary

When teachers deliberately incorporate CCCs into their teaching, there is tremendous potential to change the way students learn new content and relate it to what they already know. In doing this, children come to understand science as a way of knowing and making sense of the world around them. When teachers emphasize the crosscutting nature of energy, they help students develop a deeper understanding within and across disciplines. The Five Big Ideas serve as a common language and create meaningful touchstones, helping students develop a more integrated understanding of scientific concepts.

Teaching energy as a CCC holds both promises and challenges. These challenges include the discomfort that naturally accompanies the need to teach new standards, the lack of assessments that measure three-dimensional learning, and the absence of curriculum materials to support the teaching of CCCs. These challenges make it difficult for teachers to know how, when, and where to incorporate energy as a CCC into their teaching.

Providing high-quality PD for inservice teachers and adjusting preservice teacher curriculum to incorporate CCCs are two strategies that would help teachers incorporate CCCs into their teaching. Additionally, new assessments that emphasize the crosscutting nature of energy will provide students with opportunities to demonstrate their understanding of energy both within and across disciplines. Finally, curriculum materials that explicitly identify opportunities to make energy connections across disciplines will help teachers provide students with meaningful touchstones for deeper, more integrated learning.

Teachers and students need support from many stakeholders if they are to realize the recommendations set forth in the *Framework*. Administrators, parents, higher education faculty, and partners from business and industry all have an important role to play in supporting teachers and students as they learn and use science concepts like energy in new ways.

Conclusion

Energy plays a central role in science and society. Fittingly, energy has a prominent position in both the *Framework* and the *NGSS*. Because of its central importance in each discipline, energy has long been presented to students as a disciplinary idea, but by identifying energy as a CCC, the *Framework* emphasizes the need to rethink how we present energy across disciplines. The *Framework* stresses that CCCs are important ideas in science teaching because "they provide students with connections and intellectual tools that are related across the differing areas of disciplinary content and can enrich their application of practices and their understanding of core ideas" (NRC 2012, p. 218).

Because of its prominent position in each discipline, energy stands out as a particularly important CCC. Thus, energy provides a unique opportunity to design instruction that emphasizes the power of CCCs in science. Cary Sneider, a writing team leader for the *NGSS*, described the importance of energy in science and science teaching:

> *The same law of conservation of energy used by an engineer to design a more efficient car is used by a nutritionist to calculate the ideal meal for a patient, and by an ecologist to investigate how energy moves through an ecosystem. The crosscutting concept of energy has the potential to help students see how scientists and engineers think, and how the disciplines of biology, physics, chemistry, engineering, and Earth and space science involve similar concepts and ways of thinking. (Sneider n.d., p. 3)*

For too long, science instruction has failed to help students recognize connections between energy in living systems, physical systems, Earth systems, and everyday life.

Considering that energy is used in a variety of ways in a multitude of settings, it is remarkable that the behavior of energy can be summarized in just Five Big Ideas. By clearly and consistently presenting how these ideas are manifested in a wide range of contexts, teachers can help students develop a more robust and connected understanding of energy. If done well, science teaching that emphasizes the crosscutting nature of energy should clarify, rather than complicate, existing instruction.

Although energy is a critical concept in science and a prominent CCC, it is just one of seven CCCs identified in the *Framework*. Yet, the prominence of energy makes it a great place to start when considering how to incorporate CCCs into science instruction. The process of focusing deeply on issues related to teaching energy across grade levels and disciplines can serve as a model for emphasizing CCCs and promoting three-dimensional instruction more broadly.

Emphasizing energy as a CCC is more than simply adding another layer of discussion when learning about energy in disciplinary context; it is an intentional reframing of instruction and assessment that will help students learn about energy more efficiently and effectively across a range of settings. This is the power of a crosscutting concept in science.

References

Alberts, B. 2005. A wakeup call for science faculty. *Cell* 123 (5): 739–741.

American Association for the Advancement of Science (AAAS). 1990. *Science for all Americans*. New York: Oxford University Press.

American Association for the Advancement of Science (AAAS). 1993. *Benchmarks for science literacy*. New York: Oxford University Press.

Anderson, W. A., U. Banerjee, C. L. Drennan, S. C. R. Elgin, I. R. Epstein, J. Handelsman, G. F. Hatfull, et al. 2011. Changing the culture of science education at research universities. *Science* 331 (6014): 152–153. *http://doi.org/10.1126/science.1198280*.

Breidenstein, A., K. Fahey, C. Glickman, and F. Hensley. 2012. *Leading for powerful learning: A guide for instructional leaders*. New York: Teachers College Press.

Commission on Behavioral and Social Sciences and Education (CBASSE). 2000. *How people learn: Brain, mind, experience, and school: Expanded edition*. Washington, DC: National Academies Press.

Cooper, M. M., and M. W. Klymkowsky. 2013. The trouble with chemical energy: Why understanding bond energies requires an interdisciplinary systems approach. *Cell Biology Education* 12 (2): 306–312. *http://doi.org/10.1187/cbe.12-10-0170*.

diSessa, A., and J. F. Wagner. 2005. What coordination has to say about transfer. In *Transfer of learning from a modern multidisciplinary perspective*, ed. J. P. Mestre, 121–154. Greenwich, CT: Information Age Publishing.

Duschl, R. A., H. A. Schweingruber, and A. W. Shouse. 2007. *Taking science to school: Learning and teaching science in grades K–8*. Washington, DC: National Academies Press.

Fortus, D., and J. Krajcik. 2012. Curriculum coherence and learning progressions. In *Second international handbook of science education*, eds. B. J. Fraser, K. Tobin, and C. J. McRobbie, 783–798. Dordrecht, The Netherlands: Springer. *www.springerlink.com/index/10.1007/978-1-4020-9041-7.*

Fortus, D., L. M. Sutherland Adams, J. Krajcik, and B. Reiser. 2015. Assessing the role of curriculum coherence in student learning about energy. *Journal of Research in Science Teaching. http://doi. org/10.1002/tea.21261.*

Guskey, T. R., and K. S. Yoon. 2009. What works in professional development? *Phi Delta Kappan* March: 495–500.

Ingvarson, L., M. Meiers, and A. Beavis. 2005. Factors affecting the impact of professional development programs on teachers' knowledge, practice, student outcomes and efficacy. *Education Policy Analysis Archives* 13 (10). *http://research.acer.edu.au/professional_dev/1/.*

Institute of Medicine, National Academy of Sciences, and National Academy of Engineering. 2007. *Rising above the gathering storm: Energizing and employing America for a brighter economic future.* Washington, DC: National Academies Press.

Kali, Y., M. C. Linn, and J. E. Roseman, eds. 2008. *Designing coherent science education: Implications for curriculum, instruction, and policy.* New York: Teachers College Press.

Kesidou, S., and J. E. Roseman. 2002. How well do middle school science programs measure up? Findings from Project 2061's curriculum review. *Journal of Research in Science Teaching* 39 (6): 522–549.

Lave, J., and E. Wenger, E. 1992. *Situated learning: Legitimate peripheral participation.* Cambridge, UK: Cambridge University Press.

Levy, A., A. Scheff, R. F. Chen, P. Pelletier, and E. Fields. 2013. The BEST observation protocol: Looking at *Next Generation Science Standards'* crosscutting concepts in the classroom. Paper presented at the Annual International Meeting of the National Association for Research in Science Teaching, Rio Grande, Puerto Rico.

Linn, M. C., and B.-S. Eylon. 2000. Knowledge integration and displaced volume. *Journal of Science Education and Technology* 9 (4): 287–310.

Linn, M. C., B.-S. Eylon, E. A. Davis, and P. Bell. 2004. The knowledge integration perspective on learning. Mahwah, NJ: Lawrence Erlbaum Associates.

Loucks-Horsley, S. 1996. Professional development for science education: A critical and immediate challenge. In *National Standards & the Science Curriculum,* ed. R. Bybee, 83–90. Dubuque, IA: Kendall Hunt Publishing.

Loucks-Horsley, S., K. E. Stiles, S. E. Mundry, N. B. Love, and P. W. Hewson. 2010. *Designing professional development for teachers of science and mathematics.* 3rd ed., expanded ed. Thousand Oaks, CA: Corwin Press.

Mestre, J., ed. 2005. *Transfer of learning from a modern multidisciplinary perspective.* Greenwich, CT: Information Age Publishing.

National Research Council (NRC). 2010. *Rising above the gathering storm, revisited: Rapidly approaching Category 5.* Washington, DC: National Academies Press.

National Research Council (NRC). 2012. *A framework for K–12 science education: Practices, crosscutting concepts, and core ideas*. Washington, DC: National Academies Press.

National Research Council (NRC). 2014. *Developing assessments for the Next Generation Science Standards*. Washington, DC: National Academies Press.

NGSS Lead States. 2013. *Next Generation Science Standards: For states, by states.* Washington, DC: National Academies Press. *www.nextgenscience.org/next-generation-science-standards.*

Nordine, J., J. Krajcik, and D. Fortus. 2011. Transforming energy instruction in middle school to support integrated understanding and future learning. *Science Education* 95 (4): 670–699. *http://doi.org/10.1002/sce.20423.*

Savkar, V., and J. Lokere. 2010. *Time to decide: The ambivalence of the world of science toward education.* Cambridge, MA: Nature Education.

Sneider, C. n.d. What do I do with crosscutting concepts? *www.mheonline.com/assets/pdf/ngss/white_papers/crosscutting-concepts.pdf.*

INDEX

Page numbers printed in **boldface** type refer to tables or figures. Pages numbers followed by an *n* refer to a footnote on that page.